Direct Liquid Fuel Cells

This edited guide to direct liquid fuel cells (DLFCs) provides readers with a detailed discussion of their characteristics, mechanisms, and role in global decarbonization. Expert contributors address various facets, including their working mechanisms, sustainability, life cycle assessment, and more.

Readers will gain an expansive understanding and benefit from the discussion of various DLFCs. The contributors provide an overview of fuel cells, including their principles and role in global decarbonization, a discussion of DLFCs' sustainability and environmental impact, and a summary of the current state of knowledge on their different components, considering the importance of sustainable and low carbon emissions. They also address different liquids and their characteristics, such as methanol, ethanol, formic acid, and dimethyl ether, and the applications for which each type are best used. Encountered challenges, system design, and commercialization considerations are also discussed, with the conclusion addressing the future outlook and perspectives on DLFCs. In addition to discussing the various types of liquid fuel cells and their respective benefits and drawbacks, contributors provide real-world examples of challenging issues and their respective solutions.

This all-inclusive and cross-disciplinary resource is a vital guide for researchers and undergraduate and graduate students studying renewable energy, energy storage and conversion, and decarbonization towards net zero.

Norazuwana Shaari is a senior lecturer at Universiti Kebangsaan Malaysia and an industry consultant for UKM Pakarunding. She obtained her bachelor's and master's degrees from Universiti Sains Malaysia in 2010 and 2013, respectively, and her PhD from Universiti Kebangsaan Malaysia in 2018.

Ajaz Ahmad Wani is an assistant professor (tenure track) in the Department of Chemistry at Aligarh Muslim University (AMU), Aligarh, India. He obtained his BSc from the University of Kashmir, India, in 2011, his MSc from Jiwaji University, Gwalior, in 2014, his PhD from Aligarh Muslim University in 2022, and his postdoc from Universiti Kebangsaan Malaysia in 2024.

Siti Kartom Kamarudin has been a professor of chemical engineering at the Department of Chemical and Process Engineering, Universiti Kebangsaan Malaysia, since 2011. She is now the director of the Fuel Cell Institute of Universiti Kebangsaan Malaysia.

Direct Liquid Fuel Cells

Recent Advances and Challenges and the Sustainable Energy Transition

Edited by
Norazuwana Shaari
Ajaz Ahmad Wani
Siti Kartom Kamarudin

CRC Press
Taylor & Francis Group
Boca Raton London New York

CRC Press is an imprint of the
Taylor & Francis Group, an **informa** business

Designed cover image: Getty

First edition published 2025
by CRC Press
2385 NW Executive Center Drive, Suite 320, Boca Raton FL 33431

and by CRC Press
4 Park Square, Milton Park, Abingdon, Oxon, OX14 4RN

CRC Press is an imprint of Taylor & Francis Group, LLC

ISBN: 978-1-032-83266-1 (hbk)
ISBN: 978-1-032-84410-7 (pbk)
ISBN: 978-1-003-51259-2 (ebk)

DOI: 10.1201/9781003512592

Typeset in Palatino
by Apex CoVantage, LLC

Contents

Contributors Bios

Diya Aziz
School of Natural Sciences (SNS)
National University of Sciences & Technology (NUST)
H-12, Islamabad 44000, Pakistan
High-Temperature Energy Conversion Laboratory
Korea Institute of Energy Research
152 Gajeong-ro, Yuseong-gu, Daejeon 34129, Republic of Korea

Aabid Hussain Bhat
Department of Chemistry
National Institute of Technology
Srinagar, 190006, India

Hafsa Fatima
Department of Chemistry
Quaid- i-Azam University (QAU)
Islamabad Capital Territory 15320, Pakistan

Ram K. Gupta
Department of Chemistry
National Institute for Materials Advancement, Pittsburg State University
1701 S. Broadway, Pittsburg, KS 66762, United States

Muhammad Humayun
Department of Basic Science
University of Engineering and Technology Peshawar–25000

Nur Ain Masleeza Harun
Fuel Cell Institute
Universiti Kebangsaan Malaysia
43600 UKM Bangi, Selangor, Malaysia

Basharat Hussain
High-Temperature Energy Conversion Laboratory
Korea Institute of Energy Research
152 Gajeong-ro, Yuseong-gu, Daejeon 34129, Republic of Korea
Department of Advanced Energy and Technology
University of Science and Technology
217 Gajeong-ro, Yuseong-gu, Daejeon 34113, Republic of Korea

Siti Kartom Kamarudin
Fuel Cell Institute
Universiti Kebangsaan Malaysia
43600 UKM Bangi, Selangor, Malaysia
Department of Chemical and Process Engineering, Faculty of Engineering and Built Environment
Universiti Kebangsaan Malaysia
43600 Bangi, Selangor, Malaysia

Amjad Mumtaz Khan
Department of Chemistry
Aligarh Muslim University
Aligarh 202002, India

Muhammad Raheel Khan
Joint Doctoral School
Silesian University of Technology
Skademicka 2A, 44–100, Gliwice, Poland

Zulnurain Khan
Department of Chemistry
Aligarh Muslim University
Aligarh 202002, India

Ansari M.N.M.
Department of Mechanical Engineering, College of Engineering
Universiti Tenaga Nasional
43000 Selangor

Jayappa Manjanna
Department of Chemistry
Rani Channamma University
Belagavi—591156, Karnataka, India

Naeemakhtar Momin
Department of Chemistry
Sangolli Rayanna First Grade Constituent College
Belagavi—590017, Karnataka, India

Norsyahida Mohammad
Department of Chemical Engineering, School of Engineering and Computing
MILA University

Siti Hasanah Osman
Fuel Cell Institute
Universiti Kebangsaan Malaysia
43600 UKM Bangi, Selangor, Malaysia

Sanaullah Qamar
Department of Advanced Energy and Technology
University of Science and Technology
217 Gajeong-ro, Yuseong-gu, Daejeon 34113, Republic of Korea

Mohd Saqib
Department of Chemistry
Aligarh Muslim University
Aligarh 202002, India

Norazuwana Shaari
Fuel Cell Institute
Universiti Kebangsaan Malaysia
43600 UKM Bangi, Selangor, Malaysia

Zeba N. Siddiqui
Department of Chemistry
Aligarh Muslim University
Aligarh 202002, India

Abid Ullah
Department of Advanced Energy and Technology
University of Science and Technology
217 Gajeong-ro, Yuseong-gu, Daejeon 34113, Republic of Korea
High-Temperature Energy Conversion Laboratory
Korea Institute of Energy Research
152 Gajeong-ro, Yuseong-gu, Daejeon 34129, Republic of Korea

Ajaz Ahmad Wani
Fuel Cell Institute
Universiti Kebangsaan Malaysia
43600 UKM Bangi, Selangor, Malaysia
Department of Chemistry
Aligarh Muslim University
Aligarh 202002, India

Shariq Yousuf
Department of Chemistry
Aligarh Muslim University
Aligarh 202002, India

Zulfirdaus Zakaria
Institute for Advanced Studies
Universiti Malaya
50603, Kuala Lumpur, Malaysia

Nik Farah Hanis Nik Zaiman
Fuel Cell Institute
Universiti Kebangsaan Malaysia
43600 UKM Bangi, Selangor, Malaysia

1

Fundamental Principles of Direct Liquid Fuel Cell Technology and Role in Global Decarbonization System

Zulfirdaus Zakaria, Aabid Hussain Bhat, Shariq Yousuf, Nur Ain Masleeza Harun, Nik Farah Hanis Nik Zaiman, and Zeba N. Siddiqui

1.1 Introduction

Less reliance on fossil fuels is an urgent need in the modern world. The world's energy consumption has grown exponentially in recent years, driven by rising living standards and population density; until now, fossil fuels have been the backbone of this industry [1,2].

Decarbonizing the energy sector is essential to achieving the net-zero emission (NZE) target of the Paris Agreement, which aims to achieve this by 2050 while limiting the global temperature increase to 1.5°C. There will likely be a nearly 50% rise in the world's energy consumption by 2050 [3–6]. Without action to reduce emissions, global warming might reach 4.5–5.7°C, with levels predicted to double from preindustrial levels and a 35% rise from current levels [5,6]. By comparing the total cost of ownership of each fuel cell technology with its potential incumbent technology, the technology readiness level, and maturity in terms of the patent landscape, we analyze the technical and economic challenges in deploying fuel cells at scale, which could abate 60% of the emissions among the different energy sectors [7–9]. All energy and power production areas may benefit from fuel cells, including zero-emission transportation, backup power, and centralized power generation [10,11]. As shown in Figure 1.1a, the United Nations Conference on Trade and Development has issued a warning about the steadily rising emissions of greenhouse gases (GHG) from the global fleet of ships, with the exception of 2021, when the COVID-19 epidemic will have an impact [8,9].

DOI: 10.1201/9781003512592-1

Consequently, in order to promote the long-term viability of the shipping sector, the IMO has implemented a more stringent set of regulations, such as the International Convention for the Prevention of Pollution from Ships, the Ship Energy Efficiency Management Plan, and the Energy Efficiency Design Index, all of which seek to enhance the environmental friendliness and economic viability of ships [12,13]. The use of renewable-based fuels, particularly hydrogen and synthetic fuels produced from hydrogen, may eliminate greenhouse gas emissions entirely, as seen in Figure 1.1b, according to the IMO study [14]. Fuel cells are already an established technology, but their higher upfront and ongoing expenses are due to a lack of economies of scale and supporting infrastructure. Neither the technology nor the market is prepared to support fuel cells [10,15]. Compared to incumbent technologies, the typical TRL level of FC technology is now between five and ten, costing around five to ten times as much. Fuel cells have more initial investment costs than electric cars, but their benefits, more extended range, cheaper maintenance costs, and longer lifetimes for power generation do not outweigh them [16,17].

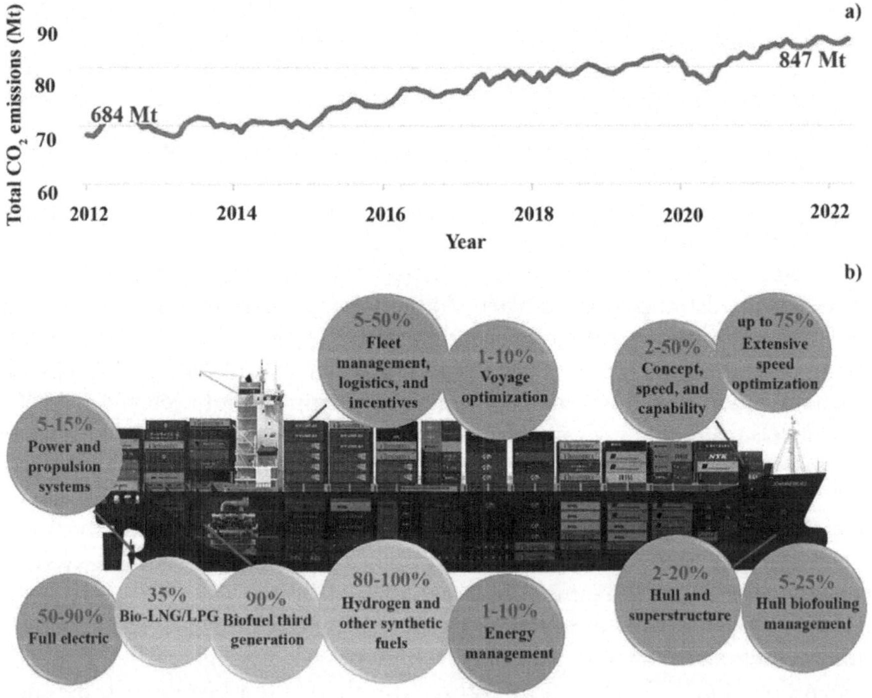

FIGURE 1.1
(a) Total CO_2 emissions from the world's maritime fleet from 2012 to 2022; (b) some solutions for cutting GHG emissions, as suggested by the IMO.

Source: Reproduced with permission from [22].

Both public and private investors should learn more about carbon-free technology. Public buildings and spaces must be fitted with emission-free technology as soon as possible to promote the technologies, develop initiatives, and generate economies of scale [7,18]. Significant economies of scale are necessary for fuel cells to achieve cost competitiveness. If customers want to be competitive and profitable in the energy sector in the future, they need to change their attention to low-carbon and hydrogen market consulting solutions. The energy industries of today won't be the same in the future [19–21].

In this chapter, we look at fuel cells, a promising technology that can potentially reduce emissions from the power and transportation industries, two of the world's most polluting industries. This chapter examines the fundamental technology of fuel cells, a principal use of hydrogen, and provides a comprehensive evaluation of several DLFCs for use in the energy industry. Energy decarbonization is an urgent need in light of rising carbon emissions; this calls for introducing carbon-free technology into the two most energy-intensive sectors, transportation and power generating. This research will shed light on fuel cell technologies, their potential uses across industries and power ranges, their relative commercial and economic feasibility, and the infrastructure and technology challenges in their widespread implementation.

1.2 Fundamental Principles of Direct Liquid Fuel Cell Technology

Since DLFCs have many benefits over fuel cells that use hydrogen-fed polymer electrolyte membranes, they are one fuel cell that is now the subject of extensive research (PEMFCs). Promising in several fields, including the automotive sector, PEMFCs utilize hydrogen as their fuel and are hailed as a ground-breaking fuel cell technology [23]. Hydrogen fuel, however, often requires storage under high pressure when in use. This is the primary concern with PEMFC safety, as it might cause explosions or other incidents involving the improper handling of hydrogen [24,25]. In addition to the issues with storage and transportation, hydrogen fuel is very combustible. Due to its liquid fuel and ease of handling, DLFCs seem to be an alternate fuel cell type [26]. More efficient power sources are required to meet the demands of today's high-tech portable electronic gadgets, including laptop computers and mobile phones, which have longer battery lives and higher power output requirements. Since liquid fuel is more convenient to handle and store, DLFC is a technology that might solve these problems [27–29]. These days, individuals rely more and more on portable electrical devices, like cellular phones and supercomputers, to prioritize their day-to-day tasks. Many industries that run on long-term, high-power generators also use these portable units. DLFC is well-suited for military instruments because it can generate power in the medium to high range, extending from 100 W to around 3 kW. Unlike DLFC, recharging lithium batteries, ubiquitous in portable electronics, require access to an electrical outlet [30,31]. Compared to lithium-ion batteries, DLFCs have a greater density (up to 600 Wh kg^{-1}), a longer operating time when used in portable devices, and are simple to recharge [29,32].

DLFCs are electrochemical devices that create electricity from the chemical energy of fuels, most often hydrogen or methanol [28,29]. The fundamental physical structure of membrane electrode assembly (MEA) in DLFC includes an electrolyte layer connected to an anode and cathode on opposite sides [33,34]. The fuel must be continually given to the anode to generate energy, while oxygen, acting as an oxidant, must be continuously supplied to the cathode [35,36]. A fuel cell's negative electrode, or anode, conducts the electrons released from the hydrogen molecules. An external circuit supplies electron to the cathode, a positive electrode [37,38]. The third main part of a fuel cell is the electrolyte, which prevents electrons from flowing through the internal circuit but allows ions to pass through it. The electrolyte is the fundamental component by which DLFC technologies are classified [39,40]. The catalyst is essential to the fuel cell since it speeds up the reaction. Here are the chemical processes that are happening at the anode and cathode [41,42]:

$$\text{Anode} H_2 \rightarrow 2H^+ + 2e^-$$
$$\text{Cathode}: {}^1/_2 O_2 + 2H^+ \rightarrow H_2O$$
$$\textbf{Overall Reaction}: \mathbf{H_2 + {}^1/_2 O_2 \rightarrow H_2O}$$

One promising technology that might contribute to hydrogen use and the goals of the Paris Agreement is fuel cells. Micropower generators, auxiliary power generators, permanent power generators, distributed power generation, portable power, transportation, military, and automotive are just a few industries that may benefit from fuel cell technology. As a result, fuel cells include all forms of electricity used in many global businesses and settings [43,44].

1.3 Types of DLFCs

The capacity to directly oxidize liquid fuels without an extra reforming process makes DLFCs a fascinating subset of fuel cell technology. Exceptional efficiency, small size, and adaptability in operation are the hallmarks of these systems. Direct fuel cells that use methanol, ethanol, or formic acid are the most common direct liquid fuel cells. However, this theme has several variations, including ones that use hydrazine or borohydride [45,46]. Some of the most researched and potentially useful varieties of DLFCs are direct methanol fuel cells (DMFCs). The anode oxidizes methanol, the principal fuel, to create carbon dioxide, protons, and electrons. At the cathode, protons combine with oxygen to form water after passing through a proton exchange membrane (PEM) [47,48]. The exceptional energy density of liquid methanol and its relative ease of handling are the defining characteristics of DMFCs. Notwithstanding these benefits, its broad use is restricted by issues such as methanol crossing the PEM and sluggish anode oxidation kinetics. The efficiency and longevity of DMFCs have been greatly aided by the creation of sophisticated catalysts, including platinum-ruthenium alloys [28,49]. A direct ethanol fuel cell (DEFC) is a double-ended fuel cell that uses ethanol instead of methanol [50,51]. There are several reasons why ethanol is preferable to methanol, but two of the most important are its lower toxicity and greater energy density. Global targets for renewable energy sources may be achieved by the sustainable production of ethanol from biomass [52,53]. In contrast, acetaldehyde and acetic acid are common byproducts of the multi-step process required to oxidize ethanol. All things considered, these intermediaries lower DEFC performance and efficiency. The complete oxidation of ethanol to carbon dioxide and the subsequent improvement of energy conversion efficiency is the primary goal of research into increasing the selectivity and catalytic activity of anode materials [54,55].

Another fascinating type of DLFC is the direct formic acid fuel cell, which uses formic acid as its fuel alone. Formic acid's many benefits are its low crossover rates, ease of handling, and high theoretical efficiency. It has a less complex oxidation route with fewer chemical intermediates than ethanol and methanol [29,56]. The material and operating expenses of DFAFCs are reduced since they generally operate at lower temperatures. However, anode catalyst poisoning, most often caused by carbon monoxide, is a significant challenge. Advancements in catalyst design have overcome these challenges, most notably with palladium-based catalysts, which allow DFAFCs to operate more efficiently and reliably [57,58]. Several DLFCs use fuels, such as hydrazine and sodium borohydride, in addition to the more common ones. Direct hydrazine fuel cells use these properties because of its high energy content and capacity to break down into nitrogen and hydrogen without forming byproducts containing carbon. This distinctive quality makes them attractive for uses necessitating a complete absence of carbon emissions. Strict handling standards are necessary because of hydrazine's toxicity and volatility, which limits its practical application [47,48].

Sodium borohydride is used in direct borohydride fuel cells because it is a high-energy-density chemical that can be hydrolyzed on-site to produce hydrogen. The cell then generates power by oxidizing the hydrogen. While borohydride's excellent theoretical efficiency is a boon to these systems, problems with water's parasitic reaction and electrode material deterioration are also obstacles. The key to solving these problems is improving electrolyte compositions and altering electrode surfaces [43,59]. There are many types of DLFC, each with its own set of properties that make it ideal for a specific task. Because of their small size and convenience of fuel storage, DMFCs and DEFCs are often preferred in portable electronic equipment. DFAFCs are useful in low-power systems that value efficiency and simplicity. Alternative fuel cell types, such as borohydride and hydrazine fuel cells, show great potential in niche markets, such as aerospace and defense, due to their low carbon emissions and high energy density [60,61].

Progress in materials science, especially in catalyst and membrane design, is crucial to develop DLFCs. Improved reaction kinetics, less poisoning effects, and longer fuel cell life are the goals of modern catalyst design. The same holds for membranes: low fuel crossover rates and strong ionic conductivity are needed for efficient functioning. There are now more ways to enhance the efficiency and dependability of DLFCs via nanotechnology and innovative materials like graphene-based composites [28,62]. Cleaner and more sustainable energy systems are becoming the norm, and DLFCs are well-positioned to play a key part in this shift. They are a foundational component of future energy technologies because of their adaptability to use various liquid fuels and the constant improvement of their design and operation. Energy generation in several areas, from consumer electronics to large-scale industrial applications, might be transformed by DLFCs by tackling present issues and capitalizing on their inherent benefits [19,23]. Regarding DLFCs and direct alcohol fuel cells (DAFCs), DMFCs are far more prevalent. There have been a lot of studies focusing on DMFC because of the great things about methanol. In its liquid state, methanol is fed rapidly into the anode, even though it is pre-reforming. With its cheap cost but high carbon energy ratio, strong electrochemical activity, biodegradability, and ease of treatment, transportation, and processing, methanol is a suitable fuel type [25,27]. While hydrogen has a relative density of 180 Wh L^{-1}, methanol has a greater density of 4820 Wh L^{-1}. Methanol, especially in fuel cell technology, requires caution due to its low toxicity. One common component of DMFC anode catalysts is a dual alloy, which is more effective than a single metal [29,63]. The anode electrochemical reaction has been enhanced by investigating several catalysts. Because of their outstanding catalytic activity towards the oxygen reduction process and the methanol oxidation reaction (MOR), platinum-based catalysts have been used on a large scale (ORR) [64,65]. To this day, PtRu remains the best option for an anode catalyst in acidic media due to its remarkable properties. Combining platinum (Pt) with ruthenium (Ru) via a supported carbon electrode may protect DMFC anodes against carbon monoxide poisoning. PtRu catalysts enhance the performance of DMFCs due to their good electrochemical reactions and increased consistency when compared to single Pt catalysts [46,62,66]. Since its discovery, several efforts have been made to synthesize PtRu-based catalysts using various techniques. The goal has been to advance the shape and structure of metals with strong electrocatalytic activity, which has been ongoing for the last 20 years. As an illustration, Xiao et al. demonstrated the synthesis of PtRu nanosponge in an acidic medium using the facile chemical reduction method with the reducing agent $NaBH_4$ as an electrocatalyst for MOR. The nanosponge exhibited a current density of 410 mA mg^{-1} Pt in 0.5 M H_2SO_4 and 1.0 M CH_3OH [29,67]. In their study, Ling et al. showed how to coat a carbon support with a poly(2,5-benzimidazole) (ABPBI) utilizing in situ polymerization. They then deposited PtRu on top of the PVP covering [68,69]. The CB/PVP/PtRu/ABPBI electrocatalyst, developed with a double-polymer coating, has been shown to have the maximum power density (103 mW cm^{-2}) and outstanding electrocatalytic activity and durability. Using magnesia (MgO) to manufacture highly alloy-unsupported PtRu DMFCs with a maximum power density of 47.18 mW cm^{-2}, Zou et al. recorded the highest power density to date [29].

Increasing the number of oxygenated species adsorption sites and increasing the activation of water by combining Pt with another metal might assist in reducing the oxidative removal of intermediate carbonaceous species (CO_{ads}) at lower potentials, reducing the adsorption of CO on Pt surfaces [64,66]. A straightforward co-reduction approach employing $NaBH_4$ at room temperature without any surfactant template was recently tried to alloy Rh metal with a Pt catalyst. This metal is rare and valuable and belongs to the same group as Ru. The unsupported PtRh electrocatalytic activity attained a current density of 404.9 mA mg^{-1} in a solution of 0.5 M H_2SO_4 and 1.0 M CH_3OH [70,71]. In a recent study, Tian et al. detailed how bimetallic PtRu catalysts enhanced catalytic activity in MOR and PtRu durability while weakening CO adsorption [65,72]. The weaker CO adsorption was confirmed by the fact that, according to the CO stripping test, the peak current potential of the CO oxidation process in NPG-Pt_4Ru_1 was ~0.52 V, which is adversely different from commercial Pt/C (0.68 V). Furthermore, in 0.5 M H_2SO_4 and 0.5 M CH_3OH, the specific current density is 2.5 times greater than that of PtRu/C (0.18 mA cm^{-2}), and it improves dramatically with an intrinsic activity value of 0.44 mA cm^{-2} when the molar ratio of Pt to Ru is 2:1 (NPG-Pt_2Ru_1) [29,72]. Compared to commercial PtRu/C, the nanoporous gold PtRu (NPG-Pt_2Ru_1) achieves a high-power density of 96 mW cm^{-2}, demonstrating a performance improvement of almost six times in active single-cell DMFCs [29,69].

1.4 Performance Optimization and Challenges

The development of DLFCs has been a giant leap forward in energy conversion, especially for fixed and mobile uses. Many things affect how well DLFCs work, including the fuel and oxidant used, the design of the cell's components, and the operating circumstances. They reduce oxygen or another oxidant at the cathode and electrochemically oxidize liquid fuels like methanol, ethanol, or formic acid. While liquid fuel storage's simplicity and high energy density promise significant benefits, several interrelated characteristics must be carefully considered to optimize their performance and overcome the problems that come with them [62,68]. The efficiency of the catalyst material in driving electrochemical processes is a crucial component determining the performance of DLFCs. The high activity and stability of platinum-based catalysts have made them the material of choice for a long time. A significant drawback is the expensive price tag and the risk of poisoning, especially from carbon monoxide byproducts in fuel oxidation. New catalysts, such as nanostructured materials, bimetallic alloys, and non-platinum group metals, have been developed to improve catalytic efficiency and poison resistance. To improve reaction kinetics, catalyst design has to advance, particularly in areas like electronic structure tuning and surface area optimization [29,67].

The fundamental component of DLFCs, membrane electrode assemblies (MEAs), are critical to the efficiency of these cells. Ionic conductivity, fuel crossover, and mechanical stability are all controlled by the choice of anion-exchange membrane (AEM) or PEM [65,67]. Although Nafion membranes are often used for PEMs, their high methanol crossover negatively affects cathode performance and fuel economy despite their strong proton conductivity. The latest advancements in membrane technology, such as functionalized polymers and composite membranes, aim to improve cell efficiency by balancing ionic conductivity with decreased fuel permeability [46,49]. To affect the performance, durability, and affordability of DLFC, the three-layer MEA membrane, gas diffusion layer (GDL), and catalyst layer (CL) are used. The MEA provides microchannels for mass transport and electrochemical processes. As shown in Figure 1.2, the MEA is structured, and water is discharged from the reactions of the DLFC by the GDL, which also supports the CL, conducts electrons, and offers gas and electron channels [50,55]. A hydrophobic material, such as carbon, water, alcohol, or polytetrafluoroethylene (PTFE), makes up the GDL [73]. When working in flood-prone environments, PTFE helps move gas and water more easily. A chemical process that transforms airborne hydrogen gas and oxygen into electrical current and water takes place in the CL. A well-dispersed catalyst with a particle size of 1–10 nm should be present inside the CL, and its thickness usually ranges from 5 to 100 μm, with a porosity of 40 to 70% [53,57].

The type and concentration of fuel are other performance determinants. Because of its high energy density and its liquid state under ambient circumstances, methanol is a preferred candidate. However, methanol crossover and slow oxidation kinetics are tricky, so fuel concentration and operations optimizations are required to overcome these challenges. Two potential alternative fuels, ethanol and formic acid, have their benefits and drawbacks. Although its reaction kinetics are slower, ethanol has a greater energy density and is less hazardous. However, formic acid may have problems with fuel stability and corrosion despite its quicker reaction kinetics [55,56].

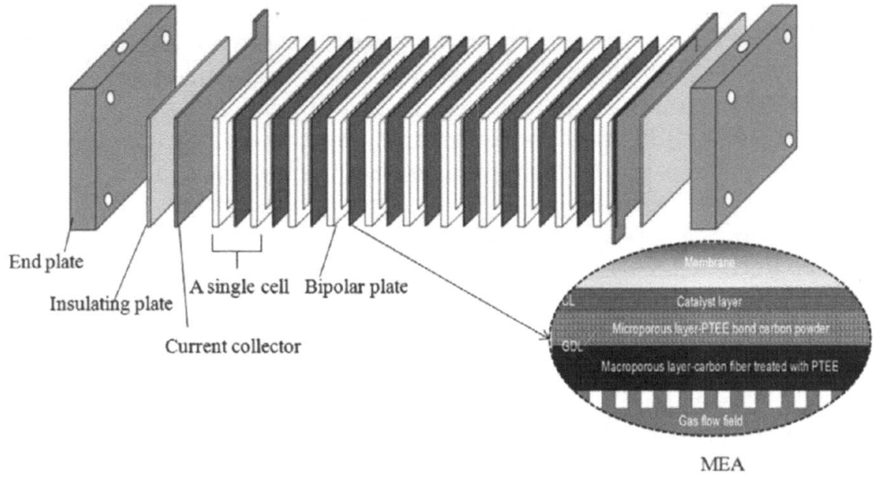

FIGURE 1.2
Basic structure of typical integrating PEMFC stack with end plate, insulating plate, current collector, bipolar plate, MEA, and the image of MEA structure.

Source: Reproduced with permission from [73].

The design of flow fields and cell structural elements significantly affect mass transfer, heat management, and water management in DLFCs. Fuel and oxidant are distributed uniformly, reaction byproducts are removed, and pressure drops are minimized by using flow fields. Improved cell performance and reactant delivery are outcomes of optimized flow field designs, like serpentine or interdigitated patterns [29,47]. Overheating may cause water to evaporate and the membrane to degrade; therefore, thermal control is essential for keeping reaction kinetics and membrane hydration at their best. Maintaining operational stability over the long term requires efficient cooling systems and heat-resistant materials. Due to its dual role as a reactant and a byproduct, water management takes on added significance in PEM-based DLFCs. When the water content of a membrane is out of whack, it might flood or dehydrate, negatively impacting gas transport and ionic conductivity. Optimized humidification methods and hydrophobic-hydrophilic balance adjustment in gas diffusion layers are two examples of the advanced tactics needed to solve these problems [29,51].

The performance of DLFCs is greatly affected by operational factors, such as reactant flow rates, temperature, and pressure. Although reaction kinetics and catalyst toxicity improve at higher temperatures, membrane stability and fuel crossover increase at such temperatures. Reducing concentration polarization and increasing efficiency, the ideal combination of pressure and flow rates guarantees an adequate supply of reactants and efficient elimination of byproducts [29,64,68]. Sophisticated control systems and real-time monitoring must balance these factors and react to changing operating conditions. There has been a lot of development, but there are still a lot of obstacles to DLFCs being widely used. Prolonged operation often results in catalyst deterioration, membrane fouling, and component corrosion; hence, durability and endurance are paramount. Creating strong materials and protective coatings that endure aggressive electrochemical conditions is key to solving these problems [28,61,63]. Fuel production, storage, and distribution network improvements are needed to make deployment economically feasible since fuel cost and infrastructure are significant obstacles. Disposal and recycling of fuel cell components are two environmental concerns that add another layer of complexity to the situation. Minimizing the ecological impact of DLFCs requires sustainable design techniques, such as using recyclable materials and ecologically benign production procedures. To further improve DLFC technology's environmental credentials, renewable energy sources like ethanol or methanol produced from biomass may be used in fuel generation [63,65].

Future research paths highlight the need to optimize systems by integrating advances in material science, cell design, and operational tactics. We may overcome constraints and broaden application options in hybrid systems by combining DLFCs with other energy storage or generation technologies, such as batteries or supercapacitors [27,74]. Developing next-generation DLFC systems may also be accelerated by using computational modelling and machine learning approaches for predictive diagnoses and performance improvement. Finally, a multidisciplinary strategy is required to optimize direct liquid fuel cells' performance, considering materials, design, and operational issues. Unlocking the full potential of DLFCs as a sustainable and efficient energy source will need continued work to address cost, durability, and environmental challenges despite substantial gains [28,75].

Fuel consumption, catalytic activity, membrane conductivity, and system integration are the four main determinants of DLFC performance optimization. Choosing the right fuel is one of the most important things affecting efficiency [28,29]. All three common fuels, methanol, ethanol, and formic acid, have their benefits and drawbacks. For example, methanol's high energy density and relative ease of handling make it a popular fuel. However, developing more sophisticated membranes and catalysts is required to address concerns, such as catalyst toxicity and methanol crossing the membrane. In contrast, ethanol is renewable and has reduced toxicity, but it needs more effective catalysts to improve performance due to its sluggish reaction kinetics and partial oxidation [28,65].

Improving DLFC efficiency still relies heavily on optimizing the catalyst. The remarkable catalytic activity of platinum-based catalysts has led to their extensive application. Their impracticality stems from their hefty price tag and vulnerability to carbon monoxide poisoning and other intermediates [27,29]. Platinum alloys and catalysts made of non-precious metals are two examples of alternative materials that scientists have been working on to overcome these obstacles. For instance, platinum-ruthenium alloys have shown enhanced resistance to carbon monoxide, and catalysts based on transition metals provide an economical option that is just as efficient [64,68,71]. Catalyst surface modification and nanostructuring are valuable ways to increase reaction rates and active surface area. It is impossible to exaggerate the importance of the membrane in DLFCs. Ion conduction and reactant separation are two of the most critical functions of PEMs, which directly affect the efficiency and lifespan of the cell [67,68]. Although conventional Nafion membranes are quite efficient in conducting protons, they are susceptible to efficiency losses due to methanol crossover. Researchers have responded by creating new membranes that are more selective while being less permeable. Composite membranes that include silica or zeolites are one example; anion-exchange membranes that function in an alkaline environment to inhibit fuel crossover and make cheaper catalysts possible are another [46,70].

Optimal performance is also heavily influenced by fuel cell design and system integration. The distribution of reactants, elimination of byproducts, and total energy production are greatly affected by the flow field design, electrode structure, and fuel delivery system. Improving efficiency has prompted research into cutting-edge microfluidic designs and 3D-printed structures for more consistent reactant distribution with less pressure loss [37,76]. Improvements at the system level are essential to keep operations stable and avoid performance deterioration over time, including appropriate ways for managing heat and water. Several obstacles prevent DLFCs from being widely used, even if they have advanced. A significant concern is how long it will last. Operating over an extended period reduces performance due to catalyst deterioration, membrane fouling, and the impacts of fuel impurities [27,72]. To address these challenges, we need to create more durable materials and diagnostic technologies that can track the causes of deterioration in real time. Furthermore, DLFC systems' scalability is still an issue. Although devices in the lab show promise, many obstacles remain to overcome before these discoveries can be used on a wide scale. These include problems with manufacturing complexity, reducing costs, and integrating with current energy infrastructures.

Resolving the DLFC technology's economic and environmental impacts is another major obstacle. The fabrication of specialized membranes and the use of precious metals as catalysts add to the high starting costs, which restricts market penetration. A few possible solutions to improve sustainability and save expenses are to recycle catalysts and membranes and to use bio-derived fuels. To ensure DLFCs contribute positively to sustainability objectives overall, assessing the environmental effects of fuel generation and cell operation is crucial. Overall, improving the efficiency of direct liquid fuel cells requires new approaches to materials, design, and system integration, among other things. Much work has gone into making things more efficient and fixing technical problems. However, there is still a way to go before we can break through the economic, scalability, and durability challenges. If these problems can be solved, DLFCs may be a game-changer in the race for a greener, more sustainable energy future [28,29,37,65].

1.5 Technological Innovations and Advances

Significant scientific advancements have characterized DLFC growth, significantly improving their efficiency, scalability, and application across several energy sectors. Improvements in reaction kinetics and reduction of overpotentials in anodic and cathodic processes have been made possible by creating high-performance catalysts. To increase their operating lives and tolerance to fuel contaminants, these catalysts, typically based on platinum and its alloys, have been subjected to severe tuning. New studies on catalysts that do not include noble metals have brought forth more affordable alternatives, which has helped to lower the financial hurdles to DLFC technology's broad adoption [10,27]. Another important area where DLFCs push the envelope is membrane technology. These ion-conducting membranes have been developed to be very resistant to chemicals, stable mechanically, and very good at transporting ions like proton or hydroxide. Research into new materials, including composite membranes and perfluorosulfonic acid (PFSA) membranes, has helped reduce performance drops caused by fuel crossover and membrane deterioration. Because of these upgrades, DLFCs have a much higher power density and last much longer when used in actual environments. Improvements in fuel technology and efficiency have also been critical in defining the current state of DLFCs [37,67]. The energy density, electrochemical reactivity, and environmental effects of several liquid fuels, including methanol and ethanol, have been the subjects of much research. One of them, ethanol, stands out as a potential contender because of its renewability and reduced toxicity compared to methanol. Scientists have been trying to find methods to improve the electro-oxidation processes of these fuels so that they can convert more energy with less waste [27,74]. Microfabrication and nanotechnology in DLFC design have opened up unseen avenues for efficiency improvement. The development of nano-engineered fuel delivery systems has allowed for a more consistent distribution of reactants, while electrodes with higher surface areas have allowed for more effective catalytic reactions. The stability of DLFC components in hostile electrochemical conditions has been improved because of advancements in nanocoatings and protective layers [28,75]. Improving thermal and water management systems has also helped DLFCs overcome problems with heat dissipation and water balancing. Optimal operating temperatures and hydration levels inside the fuel cell are achieved using modern, highly efficient heat exchangers and innovative water recovery approaches. These methods boost speed and help make DLFC systems safer and more reliable in the long run [65,68]. Intelligent control systems and automation have made DLFC operations even more revolutionary by allowing for the exact monitoring and modification of parameters like temperature, cell voltage, and fuel flow rates. DLFC management systems increasingly use AI and ML algorithms to improve operational efficiency with real-time data, provide predictive maintenance capabilities, and enhance overall system performance [77,78]. Improvements in modular design and manufacturing processes have enhanced the DLFCs' scalability. Scalable stack designs have made them suitable for larger fixed systems, while compact and lightweight variants have been created for portable applications. One game-changing technology is additive manufacturing, more commonly known as 3D printing. It allows for quickly developing and fabricating bespoke fuel cell components with complex geometries [79,80].

Several environmental factors have influenced the development of DLFC technology. The movement towards sustainable materials and production methods reflects a more significant effort to lessen the environmental impact of energy technology. The development of DLFCs is now in line with global environmental objectives, as shown by research into biodegradable membranes and the recycling of wasted catalysts [28,65]. These technical developments have broadened the range of DLFCs' potential uses, which now include stationary power production, off-grid energy solutions, transportation, and portable electronics. The development of these breakthroughs is expected to propel DLFCs to the forefront of the worldwide shift towards renewable energy sources, demonstrating the game-changing power of state-of-the-art technology in solving the energy problems of the modern day [27,37].

1.6 Applications

As a clean and efficient substitute for conventional energy sources, DLFCs have recently developed into a game-changing technology in sustainable energy. Their capacity to immediately transform the chemical energy of liquid fuels like methanol, ethanol, or formic acid into electrical energy is what makes DLFCs so versatile. Advanced power systems are among the most common uses of DLFCs. Powering electronics like laptops, military communication gear, commercial buildings, heavy vehicles, and sensors in off-grid areas is DLFC's strong suit (Figure 1.3) [81,82]. The use of DLFCs based on methanol in military field operations is an example worth mentioning. These systems provide a space-saving solution for heavy batteries, guaranteeing a constant power flow without the hassle of regular battery changes. Their stealthiness is further enhanced by their low thermal signature and silent functioning [37,78].

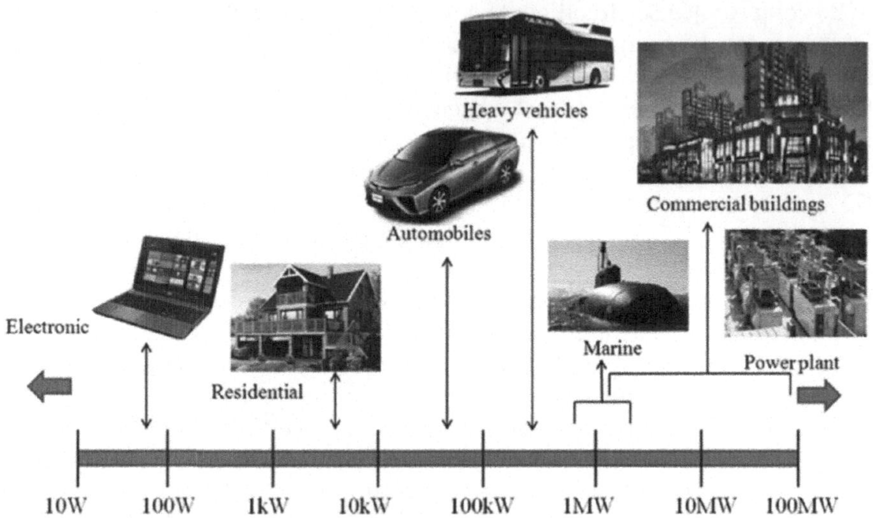

FIGURE 1.3
Scheme of some typical applications of fuel cell.

Source: Reproduced with permission from [73].

Electric vehicle propulsion systems now include DLFCs, a significant transportation sector development (EVs). Ethanol's high energy density and portability have made it an ideal fuel for DLFCs. For example, a trial run of buses fuelled by ethanol showed a substantial drop in GHG emissions when contrasted with their diesel-powered predecessors [83,84]. The environmental advantages of DLFC technology in public transportation systems were demonstrated by these buses, which not only had expanded range capabilities but also helped improve urban air quality. The use of DLFCs has also been significant in stationary power production. Backup power systems for data centres and telecommunication towers have used formic acid fuel cells. In areas vulnerable to natural catastrophes or with unstable power infrastructure, their capacity to maintain a constant power output even during grid disruptions is crucial. Their use in disaster-stricken regions to keep communication services running smoothly illustrates how resilient and flexible DLFC systems are [81,83].

The consumer electronics industry is another sector in which DLFCs have found success. Many portable electronic devices are getting their juice from small methanol fuel cells. These portable charging gadgets are a great option if you're always on the go and don't want to be tied down to an electrical outlet. Also, unlike standard lithium-ion batteries, which need a long charge, their refuelling procedure is quick, like filling up a gas tank. The use of DLFCs in this context demonstrates how they can transform how people use commonplace electronics [76,81].

Forklifts and other material-handling equipment are potential areas of use for DLFCs in the industrial sector. Forklifts driven by DLFC can be refuelled in minutes, significantly improving operating efficiency. In contrast, conventional battery-operated forklifts often have downtime owing to lengthy charging durations. Warehouse operations case studies show that DLFCs boost productivity and lessen industrial logistics' carbon footprint, which aligns with global sustainability targets [74,85]. Additionally, DLFC technology has been helpful in the healthcare sector. In remote or emergency areas, portable medical equipment that runs on DLFCs may continue to function without interruption. When there is a protracted power outage or no access to consistent electricity, devices like portable oxygen concentrators and infusion pumps with DLFCs may save lives. These applications highlight the importance of DLFCs in making healthcare systems more resilient [37,82,83].

Finally, the potential uses of DLFC are being rapidly broadened via ongoing research and development. Research initiatives are being conducted to develop hybrid energy storage and supply solutions by combining DLFCs with renewable energy systems like solar and wind. For example, there has been some success in combining solar panels with ethanol fuel cells to provide sustainable electricity for distant agricultural activities. Because it uses bioethanol produced from agricultural waste, this novel method improves energy efficiency and lends its weight to the movement towards a circular economy [27,29].

As a whole, these examples show how direct liquid fuel cells may revolutionize many different industries. With the continuous improvement of catalyst design, fuel processing, and system integration, DLFCs are becoming more and more cost-effective and efficient. This puts them in a prime position to lead the world in finding resilient and sustainable energy solutions [32,58].

A novel approach to developing electrocatalysts for alkaline direct methanol fuel cells (ADMFCs) is made possible by the high catalytic activity of non-precious metals in alkaline environments (Figure 1.4). The development of a metal-organic framework (MOF) electrocatalyst, which is loaded with NiCo non-precious metal alloy and has a surface electronic structure modulation strategy, confers resistance to CO poisoning and excellent methanol oxidation activity. The catalyst is supported by highly dispersed N-doped carbon nanofibers (CNFs). Electrocatalysts with an abundance of active sites and efficient electron transport are made possible by the combination of rapid charge transfer channels provided by the P-electron conjugated structure of polyaniline chains and porous electrospun polyacrylonitrile (PAN) nanofibers. The ADMFC single cell's anode catalyst, the NiCo/N-CNFs@800 that had been tuned, demonstrated a power density of 29.15 mW cm^{-2}. It is anticipated that NiCo/N-CNFs@800 would serve as a cheap, efficient, and CO-resistant electrocatalyst for the methanol oxidation process (MOR) because of the rapid charge and mass transfer enabled by its one-dimensional porous structure and the synergistic effect between NiCo alloy [86].

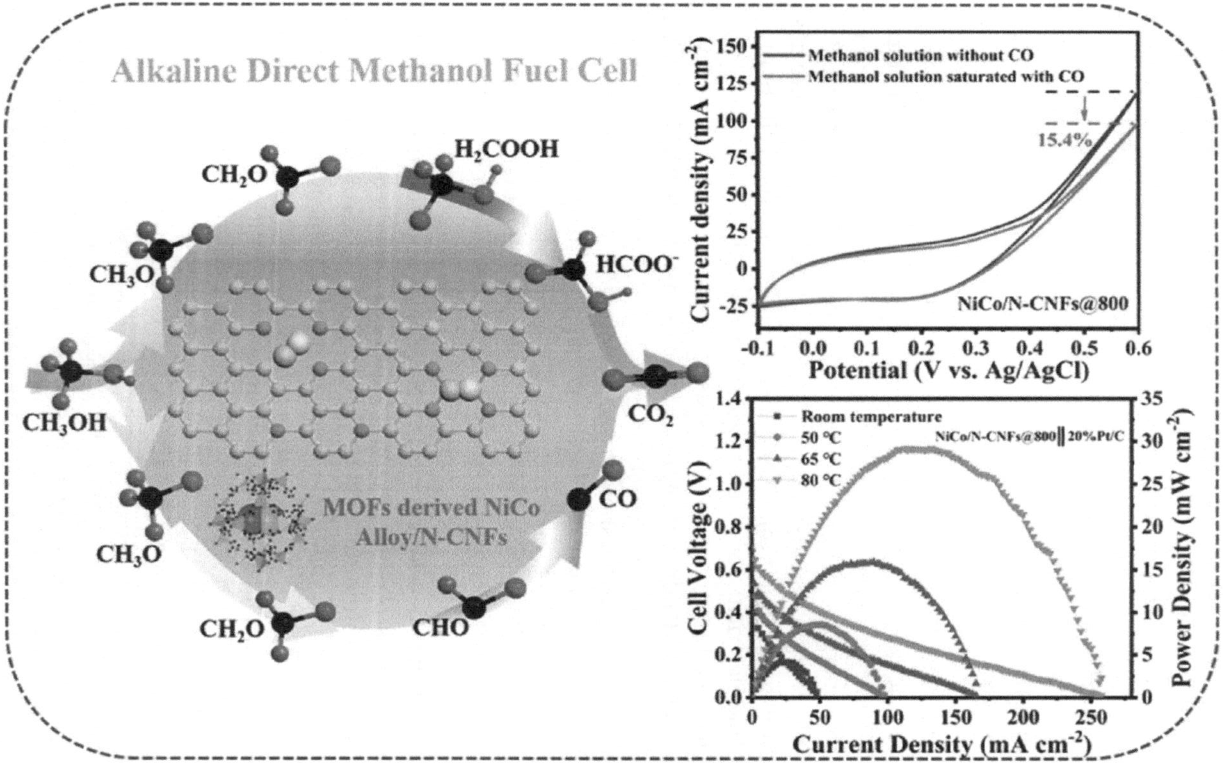

FIGURE 1.4
A metal-organic framework (MOF) based a non-precious metal alloy structure, including highly dispersed N-doped carbon nanofibers and NiCo, was created using a combination of simple mix spinning and pyrolysis treatment.

Source: Reproduced with permission from [86].

An innovative hydrazine borane ($N_2H_4BH_3$, HB) fuelling DLFC proof-of-concept has been created. The anode catalyst used in the HB electro-oxidation processes was C-PPY-Co(OH)$_2$, a non-noble, carbon-supported polypyrrole cobalt hydroxide, and it achieved an electron transfer number of 7.63 e^-. Compared to standard liquid fuels for DLFCs, HB has a higher energy density of 9947 W h kg^{-1}. The fuel cell has a greater open circuit voltage of 1.201 V compared to direct hydrazine fuel cells (DHFCs) and DMFCs. Compared to contemporary DLFCs, the fuel cell achieves a far lower crossover current density and a peak power density of 60 mW cm^{-2}. Nitrogen foam electrodes allow for a more rapid electro-oxidation of HB compared to carbon paper. Furthermore, the Ni foam electrode has a much-reduced mass transfer resistance of 54 s cm^{-1} compared to the carbon paper electrode's 231 s cm^{-1} [87].

1.7 Role in Global Decarbonization

As the world works to decarbonize, DLFCs are a game-changing technology. This method doesn't rely on conventional combustion but instead uses electrochemical processes to transform the chemical energy of fuels into electrical energy. Because of this quality, DLFCs are both ecologically friendly and very efficient, as they drastically cut down on or do away with pollution from traditional energy-generating processes, such as greenhouse gases. The main benefit of DLFCs is their adaptability to various liquid fuels, including renewable bio-alcohols, like methanol and ethanol, and other zero-carbon alternatives. Methanol is a potential fuel because it is liquid at room temperature, easily transported and stored, and has a high energy density [29,37]. Using renewable feedstocks or synthetic production methods, including collected carbon dioxide and green hydrogen, may make methanol viable for sustainable energy transportation. DLFCs that run on these fuels help close the carbon loop, which means they use recycled carbon to offset their carbon dioxide emissions. Due to its adaptability and scalability, DLFCs may be used in many applications, from small devices to massive power grids. Cleaner than traditional internal combustion engines, these fuel cells have the potential to revolutionize the transportation industry [44,88]. Also, DLFCs can help ships and planes cut down on their carbon footprints, which is excellent since they use a lot of heavy fossil fuels. When renewable energy sources, such as wind and solar, are intermittent or unavailable in a given area, DLFCs are an essential component of stationary power production. To ease the shift to a low-carbon energy infrastructure, DLFCs provide a dependable and long-term energy option [44,82].

There are fewer energy losses in DLFCs since their electrochemical processes are fundamentally more efficient than energy production based on combustion. Less fuel use is a direct result of this efficiency, which in turn helps reduce greenhouse gas emissions [27,89]. Furthermore, DLFCs have a longer lifespan and less thermal deterioration than fuel cells that use solid oxide or molten carbonate because they run at lower temperatures. Because of their lower operating temperatures, DLFCs are ideal for applications requiring rapid reaction times and quicker start-up times. The possibility of DLFC technology's seamless integration with renewable energy systems is one of its significant advantages. Using techniques like electrolysis and carbon capture, the surplus power generated by renewable energy sources may be transformed into methanol and other liquid fuels. Such fuels are subsequently used [32,44].

Despite all these benefits, DLFCs still have a way to go before they can be widely used. These obstacles include expensive manufacturing costs, components that don't last long enough, and problems with fuel crossover and catalyst poisoning. However, these constraints are being addressed by the ever-evolving field of materials science and engineering [29,32]. Improvements in fuel compositions, catalyst research, and membrane technology are making DLFC systems more efficient, stable, and cost-effective. With the ongoing improvement of these technologies, DLFCs have the potential to become an essential component of sustainable energy systems on a global scale [28,29]. In addition to being technically and environmentally sound, DLFCs have geopolitical and economic benefits since they lessen the need to import fossil fuels. Domestic liquid fuels increase energy independence and boost the local economy in countries with plenty of renewable resources [37,81]. In addition, DLFC deployment is in line with DLFCs playing a critical function in decarbonization beyond their direct uses. More people will try to rethink and reorganize the world's energy landscape if they see that alternative energy systems can work and if they encourage innovation. As an integral part of the worldwide shift to sustainable energy, DLFCs demonstrate how cutting-edge technology can meet both the short- and long-term goals of meeting energy demands while avoiding the release of carbon into the atmosphere [44,74].

1.8 Conclusion

An important step forward in sustainable energy technology is the development of DLFCs. Their functioning does not need hydrogen gas but uses liquid fuels directly, such as methanol, ethanol, or other organic molecules. Due to their efficiencies, DLFCs provide a strong case for decarbonizing the world's energy systems. They can revolutionize the energy landscapes of industries and consumers by blending into different applications. The basic idea of DLFC operation is electrochemical energy conversion, which involves the oxidation of liquid fuels at the anode to create heat, water, and electricity. This method makes the technology more accessible and safer by doing away with the complicated logistics of transporting and storing hydrogen. One advantage DLFCs have over conventional fuel cells is their capacity to run on liquid fuels, which are abundant and convenient to carry. Additionally, they are very efficient and produce minimal emissions, which perfectly harmonizes with international initiatives to promote renewable energy and decrease emissions of greenhouse gases. Decarbonization relies on DLFCs, which play an important and varied role. One option to lessen our dependence on fossil fuels is to use DLFCs, which use renewable and bio-based liquid fuels. Due to their exceptional capabilities, they are a priceless resource for attaining carbon neutrality in several industries, such as transportation, home energy, and portable power production. Their already impressive eco-friendliness is amplified by the fact that they don't undergo any carbon-intensive processing steps. Furthermore, DLFCs aid in decentralizing energy production, allowing for more distributed generation and less reliance on centralized power grids. There is tremendous hope for future energy systems in DLFC technology's development trajectory. Continuous improvements in catalyst design, efficiency of fuel use, and cell longevity are propelling their practicality and cost-effectiveness. Synthetic fuels produced from renewable energy sources are one example of the many liquid fuels being investigated as potential DLFC-compatible alternatives. These advancements might broaden the use of DLFCs, allowing them to be integrated into grid-scale energy storage systems, aircraft, and heavy-duty transport.

Promising developments in DLFC technology are anticipated due to substantial funding allocated to studying renewable energy sources. In the future, DLFCs may be able to cut costs significantly thanks to better production methods and economies of scale, according to new trends. A key component of the worldwide energy shift is anticipated to be DLFCs because of their increased fuel flexibility and conformity with circular economy concepts. Hybrid energy systems that use DLFCs are expected to provide energy networks that are both strong and resilient. These systems complement renewable energy sources like wind and solar. There could be revolutionary changes to the world's energy systems due to DLFCs. DLFCs might be instrumental in reducing the impact of climate change by easing the incorporation of renewable energy sources and offering dependable, low-carbon energy solutions. For example, the marine and aviation industries are notoriously difficult to electrify, yet they provide a workable option. On top of that, DLFCs may be scaled up or down depending on the application. Thus, they can be used in anything from tiny portable devices to massive industrial power plants. Energy fairness and reduced geopolitical conflicts over fossil fuel supplies might be achieved by the rapid adoption of DLFCs, which could hasten the shift to a sustainable energy economy. Unprecedented possibilities and difficulties await DLFCs in the future. The development of DLFCs has been a source of optimism and creativity in the face of the world's ongoing struggle with climate change and energy instability. They will play an essential part in determining the future of energy systems because of the efficient, clean, and adaptable energy solutions they offer. Continued research, regulatory backing, and industry partnership have the potential to revolutionize sustainable energy and usher in a decarbonized future via DLFCs.

References

[1] R. Avtar, S. Tripathi, A.K. Aggarwal, P. Kumar, Population–Urbanization–Energy Nexus: A Review, *Resour.* 8 (2019) 136. https://doi.org/10.3390/RESOURCES8030136.

[2] A. Kalair, N. Abas, M.S. Saleem, A.R. Kalair, N. Khan, Role of Energy Storage Systems in Energy Transition from Fossil Fuels to Renewables, *Energy Storage.* 3 (2021) e135. https://doi.org/10.1002/EST2.135.

[3] H. Moghaddasi, C. Culp, J. Vanegas, M. Ehsani, Net Zero Energy Buildings: Variations, Clarifications, and Requirements in Response to the Paris Agreement, *Energies.* 14 (2021) 3760. https://doi.org/10.3390/EN14133760.

[4] H.C. Lau, S.C. Tsai, Global Decarbonization: Current Status and What It Will Take to Achieve Net Zero by 2050, *Energies.* 16 (2023) 7800. https://doi.org/10.3390/EN16237800.

[5] D.S. Renné, Progress, Opportunities and Challenges of Achieving Net-Zero Emissions and 100% Renewables, *Sol. Compass.* 1 (2022) 100007. https://doi.org/10.1016/J.SOLCOM.2022.100007.

[6] O.C. Anika, S.G. Nnabuife, A. Bello, E.R. Okoroafor, B. Kuang, R. Villa, Prospects of Low and Zero-Carbon Renewable Fuels in 1.5-Degree Net Zero Emission Actualisation by 2050: A Critical Review, *Carbon Capture Sci. Technol.* 5 (2022) 100072. https://doi.org/10.1016/J.CCST.2022.100072.

[7] H.N. Khokhar, Decarbonisation of Global Economies; Is Net Zero Emission Achievable? The Case for Hydrogen and Fuel Cell Technology for Innovative Futures, *J. Entrep. Innov. Emerg. Econ.* 9 (2023) 92–130. https://doi.org/10.1177/23939575221141578/ASSET/IMAGES/LARGE/10.1177_23939575221141578-FIG20.JPEG.

[8] A. Gambhir, R. Green, M. Grubb, P. Heptonstall, C. Wilson, R. Gross, How Are Future Energy Technology Costs Estimated? Can We Do Better? *Int. Rev. Environ. Resour. Econ.* 15 (2021) 271–318. https://doi.org/10.1561/101.00000128.

[9] Y. Zhu, J. Wu, O. Gaidai, Y. Zhu, J. Wu, O. Gaidai, The Technology Innovation of Hybrid Electric Vehicles: A Patent-Based Study, *World Electr. Veh. J.* 15 (2024) 329. https://doi.org/10.3390/WEVJ15080329.

[10] I. Staffell, D. Scamman, A. Velazquez Abad, P. Balcombe, P.E. Dodds, P. Ekins, N. Shah, K.R. Ward, The Role of Hydrogen and Fuel Cells in the Global Energy System, *Energy Environ. Sci.* 12 (2019) 463–491. https://doi.org/10.1039/C8EE01157E.

[11] M. Mac Kinnon, G. Razeghi, S. Samuelsen, The Role of Fuel Cells in Port Microgrids to Support Sustainable Goods Movement, Renew. *Sustain. Energy Rev.* 147 (2021) 111226. https://doi.org/10.1016/J.RSER.2021.111226.

[12] M. Issa, A. Ilinca, F. Martini, Ship Energy Efficiency and Maritime Sector Initiatives to Reduce Carbon Emissions, *Energies.* 15 (2022) 7910. https://doi.org/10.3390/EN15217910.
[13] P. Serra, G. Fancello, Towards the IMO's GHG Goals: A Critical Overview of the Perspectives and Challenges of the Main Options for Decarbonizing International Shipping, *Sustain.* 12 (2020) 3220. https://doi.org/10.3390/SU12083220.
[14] A.T. Hoang, A. Pandey, F.J. Martinez De Osés, W.H. Chen, Z. Said, K.H. Ng, Ü. Ağbulut, W. Tarełko, A.I. Ölçer, X.P. Nguyen, Technological Solutions for Boosting Hydrogen Role in Decarbonization Strategies and Net-Zero Goals of World Shipping: Challenges and Perspectives, *Renew. Sustain. Energy Rev.* 188 (2023) 113790. https://doi.org/10.1016/J.RSER.2023.113790.
[15] D.L. Greene, J.M. Ogden, Z. Lin, Challenges in the Designing, Planning and Deployment of Hydrogen Refueling Infrastructure for Fuel Cell Electric Vehicles, *ETransportation.* 6 (2020) 100086. https://doi.org/10.1016/J.ETRAN.2020.100086.
[16] A. Ajanovic, R. Haas, Economic and Environmental Prospects for Battery Electric- and Fuel Cell Vehicles: A Review, *Fuel Cells.* 19 (2019) 515–529. https://doi.org/10.1002/FUCE.201800171.
[17] Z.P. Cano, D. Banham, S. Ye, A. Hintennach, J. Lu, M. Fowler, Z. Chen, Batteries and Fuel Cells for Emerging Electric Vehicle Markets, *Nat. Energy.* 3 (2018) 279–289. https://doi.org/10.1038/s41560-018-0108-1.
[18] S. Harichandan, S.K. Kar, P.K. Rai, A Systematic and Critical Review of Green Hydrogen Economy in India, *Int. J. Hydrogen Energy.* 48 (2023) 31425–31442. https://doi.org/10.1016/J.IJHYDENE.2023.04.316.
[19] D. Xu, Z. Liu, R. Shan, H. Weng, H. Zhang, A. Meneghetti, P. Simeoni, D. Xu, Z. Liu, R. Shan, H. Weng, H. Zhang, How a Grid Company Could Enter the Hydrogen Industry through a New Business Model: A Case Study in China, *Sustain.* 15 (2023) 4417. https://doi.org/10.3390/SU15054417.
[20] C. Mazur, S. Hall, J. Hardy, M. Workman, Technology Is Not a Barrier: A Survey of Energy System Technologies Required for Innovative Electricity Business Models Driving the Low Carbon Energy Revolution, *Energies.* 12 (2019) 428. https://doi.org/10.3390/EN12030428.
[21] A. Sułek, P.F. Borowski, Business Models on the Energy Market in the Era of a Low-Emission Economy, *Energies.* 17 (2024) 3235. https://doi.org/10.3390/EN17133235.
[22] A.T. Hoang, A. Pandey, F.J. Martinez De Osés, W.H. Chen, Z. Said, K.H. Ng, Ü. Ağbulut, W. Tarełko, A.I. Ölçer, X.P. Nguyen, Technological Solutions for Boosting Hydrogen Role in Decarbonization Strategies and Net-Zero Goals of World Shipping: Challenges and Perspectives, *Renew. Sustain. Energy Rev.* 188 (2023) 113790. https://doi.org/10.1016/J.RSER.2023.113790.
[23] T. Wilberforce, A.G. Olabi, I. Muhammad, A. Alaswad, E.T. Sayed, A.G. Abo-Khalil, H.M. Maghrabie, K. Elsaid, M.A. Abdelkareem, Recovery of Waste Heat from Proton Exchange Membrane Fuel Cells—A Review, *Int. J. Hydrogen Energy.* 52 (2024) 933–972. https://doi.org/10.1016/J.IJHYDENE.2022.08.069.
[24] J. Hong, J. Yang, Z. Weng, F. Ma, F. Liang, C. Zhang, Review on Proton Exchange Membrane Fuel Cells: Safety Analysis and Fault Diagnosis, *J. Power Sources.* 617 (2024) 235118. https://doi.org/10.1016/J.JPOWSOUR.2024.235118.
[25] Y.-Q. Sun, Explosions of Hydrogen Storages and the Safety Considerations in Hydrogen-Powered Railway Applications—A Review, *Hydrog.* 5 (2024) 901–918. https://doi.org/10.3390/HYDROGEN5040047.

[26] H. Çelık Kazici, Ş. Yilmaz, T. Şahan, F. Yildiz, Ö. F. Er, H. Kivrak, A Comprehensive Study of Hydrogen Production from Ammonia Borane via PdCoAg/AC Nanoparticles and Anodic Current in Alkaline Medium: Experimental Design with Response Surface Methodology, *Front. Energy.* 14 (2020) 578–589. https://doi.org/10.1007/S11708-020-0808-7/METRICS.

[27] F.M. Souza, V.S. Pinheiro, T.C. Gentil, L.E.B. Lucchetti, J.C.M. Silva, M.L.M.G. Santos, I. De Oliveira, W.M.C. Dourado, G. Amaral-Labat, S. Okamoto, M.C. Santos, Alkaline Direct Liquid Fuel Cells: Advances, Challenges and Perspectives, *J. Electroanal. Chem.* 922 (2022) 116712. https://doi.org/10.1016/J.JELECHEM.2022.116712.

[28] S.H. Osman, O.H. Chyuan, S.K. Kamarudin, N. Shaari, I.H. Hanapi, Z. Zakaria, N.H. Ahmad Zaidi, S.A. Adnan, R. Bahru, Nanocatalysts in Direct Liquid Fuel Cells: Advancements for Superior Performance and Energy Sustainability, *Int. J. Green Energy.* (2024). https://doi.org/10.1080/15435075.2024.2396070.

[29] N. Shaari, S.K. Kamarudin, R. Bahru, S.H. Osman, N.A.I. Md Ishak, Progress and Challenges: Review for Direct Liquid Fuel Cell, *Int. J. Energy Res.* 45 (2021) 6644–6688. https://doi.org/10.1002/ER.6353.

[30] J.A. Nogueira, H. Varela, Direct Liquid Fuel Cells—The Influence of Temperature and Dynamic Instabilities, *Energy Fuels.* 34 (2020) 12995–13009. https://doi.org/10.1021/ACS.ENERGYFUELS.0C02491/SUPPL_FILE/EF0C02491_SI_001.PDF.

[31] M. Tanveer, T. Ambreen, H. Khan, G.M. Kim, C.W. Park, Paper-Based Microfluidic Fuel Cells and Their Applications: A Prospective Review, *Energy Convers. Manag.* 264 (2022) 115732. https://doi.org/10.1016/j.enconman.2022.115732.

[32] Ş. Neaţu, F. Neaţu, I. M. Chirica, I. Borbáth, E. Tálas, A. Tompos, S. Somacescu, P. Osiceanu, M. A. Folgado, A. M. Chaparro, M. Florea, Recent Progress in Electrocatalysts and Electrodes for Portable Fuel Cells, *J. Mater. Chem. A.* 9 (2021) 17065–17128. https://doi.org/10.1039/D1TA03644K.

[33] T. Krauskopf, F.H. Richter, W.G. Zeier, J. Janek, Physicochemical Concepts of the Lithium Metal Anode in Solid-State Batteries, *Chem. Rev.* 120 (2020) 7745–7794. https://doi.org/10.1021/ACS.CHEMREV.0C00431/SUPPL_FILE/CR0C00431_SI_001.PDF.

[34] A. Manthiram, A Reflection on Lithium-Ion Battery Cathode Chemistry, *Nat. Commun.* 11 (2020) 1–9. https://doi.org/10.1038/s41467-020-15355-0.

[35] M.A. Abdulwahhab, S.T. Najim, M.A. Abdulwahhab, Microbial Fuel Cell Review: Thermodynamic and Infuencing Factors, Exoelectrogenic Bacteria, Anode and Cathode Configuration, *J. Chem. Technol. Biotechnol.* 98 (2023) 1559–1583. https://doi.org/10.1002/JCTB.7371.

[36] M. Ramya, P. Senthil Kumar, A Review on Recent Advancements in Bioenergy Production Using Microbial Fuel Cells, *Chemosphere.* 288 (2022) 132512. https://doi.org/10.1016/J.CHEMOSPHERE.2021.132512.

[37] N. Sazali, W.N.W. Salleh, A.S. Jamaludin, M.N.M. Razali, New Perspectives on Fuel Cell Technology: A Brief Review, *Membranes.* 10 (2020) 99. https://doi.org/10.3390/MEMBRANES10050099.

[38] L. Fan, Z. Tu, S.H. Chan, Recent Development of Hydrogen and Fuel Cell Technologies: A Review, *Energy Rep.* 7 (2021) 8421–8446. https://doi.org/10.1016/J.EGYR.2021.08.003.

[39] Y. Manoharan, S.E. Hosseini, B. Butler, H. Alzhahrani, B.T.F. Senior, T. Ashuri, J. Krohn, Hydrogen Fuel Cell Vehicles; Current Status and Future Prospect, *Appl. Sci.* 9 (2019) 2296. https://doi.org/10.3390/APP9112296.

[40] D. Wu, C. Peng, C. Yin, H. Tang, Review of System Integration and Control of Proton Exchange Membrane Fuel Cells, *Electrochem. Energy Rev.* 3 (2020) 466–505. https://doi.org/10.1007/S41918-020-00068-1/METRICS.
[41] Z. Jia, X. Lin, C. Li, A Review of Fuel Cell Cathode Catalysts Based on Hollow Porous Materials for Improving Oxygen Reduction Performance, *Catal. Sci. Technol.* 14 (2024) 5505–5524. https://doi.org/10.1039/D4CY00830H.
[42] N. Hassan, Catalytic Performance of Nanostructured Materials Recently Used for Developing Fuel Cells' Electrodes, *Int. J. Hydrogen Energy.* 46 (2021) 39315–39368. https://doi.org/10.1016/J.IJHYDENE.2021.09.177.
[43] Y. Luo, Y. Wu, B. Li, T. Mo, Y. Li, S.P. Feng, J. Qu, P.K. Chu, Development and Application of Fuel Cells in the Automobile Industry, *J. Energy Storage.* 42 (2021) 103124. https://doi.org/10.1016/J.EST.2021.103124.
[44] A.G. Olabi, T. Wilberforce, M.A. Abdelkareem, Fuel Cell Application in the Automotive Industry and Future Perspective, *Energy.* 214 (2021) 118955. https://doi.org/10.1016/J.ENERGY.2020.118955.
[45] A. Kumar, P. Daw, D. Milstein, Homogeneous Catalysis for Sustainable Energy: Hydrogen and Methanol Economies, Fuels from Biomass, and Related Topics, *Chem. Rev.* 122 (2022) 385–441. https://doi.org/10.1021/ACS.CHEMREV.1C00412/ASSET/IMAGES/MEDIUM/CR1C00412_0046.GIF.
[46] Z.A.C. Ramli, N. Shaari, T.S.T. Saharuddin, Progress and Major BARRIERS of Nanocatalyst Development in Direct Methanol Fuel Cell: A Review, *Int. J. Hydrogen Energy.* 47 (2022) 22114–22146. https://doi.org/10.1016/J.IJHYDENE.2022.04.239.
[47] Y. Zhou, H. Zhong, S. Chen, G. Wen, L. Shen, Y. Wang, R. Chen, L. Tao, S. Wang, Proton Exchange Membrane-Based Electrocatalytic Systems for Hydrogen Production, *Carbon Energy.* (2024) e629. https://doi.org/10.1002/CEY2.629.
[48] A. Yuda, A. Ashok, A. Kumar, A Comprehensive and Critical Review on Recent Progress in Anode Catalyst for Methanol Oxidation Reaction, *Catal. Rev.* 64 (2022) 126–228. https://doi.org/10.1080/01614940.2020.1802811.
[49] P. Agrawal, S. Ebrahim, D. Ponnamma, Advancements in Nanocarbon-Based Catalysts for Enhanced Fuel Cell Performance: A Comprehensive Review, *Int. J. Energy Water Resour.* (2024) 1–23. https://doi.org/10.1007/S42108-024-00324-W/FIGURES/13.
[50] A. Fan, J. Wang, Y. He, M. Perčić, N. Vladimir, L. Yang, Decarbonising Inland Ship Power System: Alternative Solution and Assessment Method, *Energy.* 226 (2021) 120266. https://doi.org/10.1016/J.ENERGY.2021.120266.
[51] M. Perčić, N. Vladimir, A. Fan, Techno-Economic Assessment of Alternative Marine Fuels for Inland Shipping in Croatia, *Renew. Sustain. Energy Rev.* 148 (2021) 111363. https://doi.org/10.1016/J.RSER.2021.111363.
[52] M. Antar, D. Lyu, M. Nazari, A. Shah, X. Zhou, D.L. Smith, Biomass for a Sustainable Bioeconomy: An Overview of World Biomass Production and Utilization, *Renew. Sustain. Energy Rev.* 139 (2021) 110691. https://doi.org/10.1016/J.RSER.2020.110691.
[53] N.M. Clauser, G. González, C.M. Mendieta, J. Kruyeniski, M.C. Area, M.E. Vallejos, Biomass Waste as Sustainable Raw Material for Energy and Fuels, *Sustain.* 13 (2021) 794. https://doi.org/10.3390/SU13020794.
[54] Y. Zheng, X. Wan, X. Cheng, K. Cheng, Z. Liu, Z. Dai, Advanced Catalytic Materials for Ethanol Oxidation in Direct Ethanol Fuel Cells, *Catalysts.* 10 (2020) 166. https://doi.org/10.3390/CATAL10020166.

[55] D. Tang, G. Lu, Z. Shen, Y. Hu, L. Yao, B. Li, G. Zhao, B. Peng, X. Huang, A Review on Photo-, Electro- and Photoelectro- Catalytic Strategies for Selective Oxidation of Alcohols, *J. Energy Chem.* 77 (2023) 80–118. https://doi.org/10.1016/J.JECHEM.2022.10.038.
[56] P.N.A.M. Othman, N.A. Karim, S.K. Kamarudin, First Principle Study of the Electronic and Catalytic Properties of Palladium-Silver (PdAg) Alloys Catalyst for Direct Liquid Fuel Cells, *Chem. Phys.* 564 (2023) 111711. https://doi.org/10.1016/J.CHEMPHYS.2022.111711.
[57] H. Cheng, J. Wang, C. Wu, Z. Liu, Electrocatalysts for Formic Acid-Powered PEM Fuel Cells: Challenges and Prospects, *Energy Mater. Adv.* (2023). https://doi.org/10.34133/ENERGYMATADV.0067/ASSET/38987416-ED76-4704-B39F-AA56F772AD3F/ASSETS/GRAPHIC/ENERGYMATADV.0067.FIG.012.JPG.
[58] S. Mohanapriya, D. Gopi, Electro-Oxidation of Alcohols—Recent Advancements in Synthesis and Applications of Palladium Core-Shell Nanostructured Model Catalysts, *Renew. Sustain. Energy Rev.* 148 (2021) 111211. https://doi.org/10.1016/J.RSER.2021.111211.
[59] L. Ouyang, J. Jiang, K. Chen, M. Zhu, Z. Liu, Hydrogen Production via Hydrolysis and Alcoholysis of Light Metal-Based Materials: A Review, *Nano-Micro Lett.* 13 (2021) 1–30. https://doi.org/10.1007/S40820-021-00657-9.
[60] N. Bisht, P. More, P.K. Khanna, R. Abolhassani, Y.K. Mishra, M. Madsen, Progress of Hybrid Nanocomposite Materials for Thermoelectric Applications, *Mater. Adv.* 2 (2021) 1927–1956. https://doi.org/10.1039/D0MA01030H.
[61] B.C. Tashie-Lewis, S.G. Nnabuife, Hydrogen Production, Distribution, Storage and Power Conversion in a Hydrogen Economy—A Technology Review, *Chem. Eng. J. Adv.* 8 (2021) 100172. https://doi.org/10.1016/J.CEJA.2021.100172.
[62] A. Kaur, G. Kaur, P.P. Singh, S. Kaushal, Supported Bimetallic Nanoparticles as Anode Catalysts for Direct Methanol Fuel Cells: A Review, *Int. J. Hydrogen Energy.* 46 (2021) 15820–15849. https://doi.org/10.1016/J.IJHYDENE.2021.02.037.
[63] D.M. Fadzillah, S.K. Kamarudin, M.A. Zainoodin, M.S. Masdar, Critical Challenges in the System Development of Direct Alcohol Fuel Cells as Portable Power Supplies: An Overview, *Int. J. Hydrogen Energy.* 44 (2019) 3031–3054. https://doi.org/10.1016/J.IJHYDENE.2018.11.089.
[64] Z.A.C. Ramli, S.K. Kamarudin, Platinum-Based Catalysts on Various Carbon Supports and Conducting Polymers for Direct Methanol Fuel Cell Applications: A Review, *Nanoscale Res. Lett.* 13 (2018) 1–25. https://doi.org/10.1186/S11671-018-2799-4.
[65] H. Tian, Y. Yu, Q. Wang, J. Li, P. Rao, R. Li, Y. Du, C. Jia, J. Luo, P. Deng, Y. Shen, X. Tian, Recent Advances in Two-Dimensional Pt Based Electrocatalysts for Methanol Oxidation Reaction, *Int. J. Hydrogen Energy.* 46 (2021) 31202–31215. https://doi.org/10.1016/J.IJHYDENE.2021.07.006.
[66] S. Bong, D. Han, Supercritical, Freezing and Thermal Drying Process of Resorcinol-Formaldehyde Polymer Based Nano-Carbons and Their Highly Loaded PtRu Anode Electrocatalyst for DMFC, *Electroanalysis.* 31 (2019) 1311–1315. https://doi.org/10.1002/ELAN.201900013.
[67] L. Gong, Z. Yang, K. Li, W. Xing, C. Liu, J. Ge, Recent Development of Methanol Electrooxidation Catalysts for Direct Methanol Fuel Cell, *J. Energy Chem.* 27 (2018) 1618–1628. https://doi.org/10.1016/J.JECHEM.2018.01.029.

[68] M.S. Khalifeh-Soltani, E. Shams, E. Sharifi, Pt–Ru Nanoparticles Anchored on Poly(Brilliant Cresyl Blue) as a New Polymeric Support: Application as an Efficient Electrocatalyst in Methanol Oxidation Reaction, *Int. J. Hydrogen Energy.* 45 (2020) 849–860. https://doi.org/10.1016/J.IJHYDENE.2019.10.071.

[69] S.H. Osman, S.K. Kamarudin, N.A. Karim, S. Basri, Application of Graphene in Low-Temperature Fuel Cell Technology: An Overview, *Int. J. Energy Res.* 45 (2021) 18318–18336. https://doi.org/10.1002/ER.6969.

[70] B.A. Yusuf, W. Yaseen, J. Xie, A.A. Babangida, A.I. Muhammad, M. Xie, Y. Xu, Rational Design of Noble Metal-Based Multimetallic Nanomaterials: A Review, *Nano Energy.* 104 (2022) 107959. https://doi.org/10.1016/J.NANOEN.2022.107959.

[71] S. Ashraf, Y. Liu, H. Wei, R. Shen, H. Zhang, X. Wu, S. Mehdi, T. Liu, B. Li, Bimetallic Nanoalloy Catalysts for Green Energy Production: Advances in Synthesis Routes and Characterization Techniques, *Small.* 19 (2023) 2303031. https://doi.org/10.1002/SMLL.202303031.

[72] K. Wang, T. Zhou, J. He, Z. Cao, Z. Jiang, Regulating the Atomic Ratio of Pt/Ru to Enhance CO Anti-Poisoning of Pt Based Electrocatalysts Toward Methanol Oxidation Reaction, *Mol. Catal.* 556 (2024) 113927. https://doi.org/10.1016/J.MCAT.2024.113927.

[73] L. Fan, Z. Tu, S.H. Chan, Recent Development of Hydrogen and Fuel Cell Technologies: A Review, *Energy Rep.* 7 (2021) 8421–8446. https://doi.org/10.1016/J.EGYR.2021.08.003.

[74] J.A. Gómez, D.M.F. Santos, R. Salkuti, D. Zhu, J.A. Gómez, D.M.F. Santos, The Status of On-Board Hydrogen Storage in Fuel Cell Electric Vehicles, *Designs.* 7 (2023) 97. https://doi.org/10.3390/DESIGNS7040097.

[75] R. Rajalakshmi, N. Ponpandian, Carbon Nanostructures for Flexible and Lightweight Fuel Cells, *Handb. Funct. Carbon Nanostructures.* (2024) 1–34. https://doi.org/10.1007/978-3-031-14955-9_45-1.

[76] G. Zhang, Z. Qu, W.Q. Tao, X. Wang, L. Wu, S. Wu, X. Xie, C. Tongsh, W. Huo, Z. Bao, K. Jiao, Y. Wang, Porous Flow Field for Next-Generation Proton Exchange Membrane Fuel Cells: Materials, Characterization, Design, and Challenges, *Chem. Rev.* 123 (2023) 989–1039. https://doi.org/10.1021/ACS.CHEMREV.2C00539/ASSET/IMAGES/MEDIUM/CR2C00539_0023.GIF.

[77] L. Yang, J. Driscol, S. Sarigai, Q. Wu, C.D. Lippitt, M. Morgan, Towards Synoptic Water Monitoring Systems: A Review of AI Methods for Automating Water Body Detection and Water Quality Monitoring Using Remote Sensing, *Sensors.* 22 (2022) 2416. https://doi.org/10.3390/S22062416.

[78] L. Jia, Z. Li, Z. Hu, Applications of the Internet of Things in Renewable Power Systems: A Survey, *Energies.* 17 (2024) 4160. https://doi.org/10.3390/EN17164160.

[79] I.M. Alarifi, Revolutionising Fabrication Advances and Applications of 3D Printing with Composite Materials: A Review, *Virtual Phys. Prototyp.* 19 (2024). https://doi.org/10.1080/17452759.2024.2390504.

[80] M.A. Islam, M.H. Mobarak, M.I.H. Rimon, M.Z. Al Mahmud, J. Ghosh, M.M.S. Ahmed, N. Hossain, Additive Manufacturing in Polymer Research: Advances, Synthesis, and Applications, *Polym. Test.* 132 (2024) 108364. https://doi.org/10.1016/J.POLYMERTESTING.2024.108364.

[81] N.A.A. Qasem, G.A.Q. Abdulrahman, A Recent Comprehensive Review of Fuel Cells: History, Types, and Applications, *Int. J. Energy Res.* 2024 (2024) 7271748. https://doi.org/10.1155/2024/7271748.

[82] R. Rath, P. Kumar, S. Mohanty, S.K. Nayak, Recent Advances, Unsolved Deficiencies, and Future Perspectives of Hydrogen Fuel Cells in Transportation and Portable Sectors, *Int. J. Energy Res.* 43 (2019) 8931–8955. https://doi.org/10.1002/ER.4795.
[83] Ş. Altun, M.Ş. Adin, K. İlçin, Monohydric Aliphatic Alcohols as Liquid Fuels for Using in Internal Combustion Engines: A Review. 238 (2023) 1941–1975. https://doi.org/10.1177/09544089231160472.
[84] J. Forsberg, A. Krook-riekkola, Recoupling Climate Change and Air Quality: Exploring Low-Emission Options in Urban Transportation Using the TIMES-City Model, *Energies.* 14 (2021) 3220. https://doi.org/10.3390/EN14113220.
[85] L.C. Tomé, D.M.F. Santos, S. Velizarov, I.M. Coelhoso, A. Mendes, J.G. Crespo, M.N. de Pinho, Overview of Membrane Science and Technology in Portugal, *Membranes.* 12 (2022) 197. https://doi.org/10.3390/MEMBRANES12020197.
[86] F. Chen, S. Guo, S. Yu, C. Zhang, M. Guo, C. Li, Hierarchical N-Doped Carbon Nanofiber-Loaded NiCo Alloy Nanocrystals with Enhanced Methanol Electrooxidation for Alkaline Direct Methanol Fuel Cells, *J. Colloid Interface Sci.* 646 (2023) 43–53. https://doi.org/10.1016/J.JCIS.2023.05.030.
[87] Y. Zhang, G. Zhu, Z. Chen, Y. Liu, D. Ye, A. Wang, W. Jiang, C. Deng, X. Zhuang, J. Zhang, C. Ke, Direct Hydrazine Borane Fuel Cells Using Non-Noble Carbon-Supported Polypyrrole Cobalt Hydroxide as an Anode Catalyst, *Sustain. Energy Fuels.* 7 (2023) 2594–2600. https://doi.org/10.1039/D3SE00329A.
[88] N. Duarte Souza Alvarenga Santos, V. Rückert Roso, A.C. Teixeira Malaquias, J.G. Coelho Baêta, Internal Combustion Engines and Biofuels: Examining Why This Robust Combination Should Not Be Ignored for Future Sustainable Transportation, *Renew. Sustain. Energy Rev.* 148 (2021) 111292. https://doi.org/10.1016/J.RSER.2021.111292.
[89] Z. Ma, U. Legrand, E. Pahija, J.R. Tavares, D.C. Boffito, From CO2to Formic Acid Fuel Cells, *Ind. Eng. Chem. Res.* 60 (2021) 803–815. https://doi.org/10.1021/ACS.IECR.0C04711/ASSET/IMAGES/MEDIUM/IE0C04711_0008.GIF.

2

System Design, Sustainability, and Efficiency in Direct Liquid Fuel Cells

Diya Aziz, Sanaullah Qamar, Hafsa Fatima, Abid Ullah, and Basharat Hussain

2.1 Introduction

The world is focusing on cutting-edge technologies that have the potential to completely change the energy landscape in the search for sustainable energy solutions. Direct liquid fuel cells are one of the most intriguing alternatives among them; they have the potential to completely change the way we produce and use energy. Using less emissions and more flexibility than conventional energy sources, these amazing electrochemical gadgets efficiently generate electricity by utilizing the energy contained in liquid fuels [1]. DLFCs have their origins in the ground-breaking work of researchers seeking to surpass the constraints of conventional fuel cells. Unlike their counterparts that rely on hydrogen as the main fuel, DLFCs possess the remarkable ability to directly harness a diverse array of liquid fuels, including methanol, ethanol, formic acid, and even bio-derived solutions. This versatility not only broadens the potential applications but also addresses challenges associated with fuel storage, distribution, and infrastructure [2]. The importance of system design and efficiency in DLFCs cannot be overstated, as these components are necessary to fully realize the potential of this ground-breaking technology. System design refers to the intricate relationships between several components that affect the overall viability and performance of DLFCs, such as fuel supply systems, catalyst design, and cell layout. Apart from maximizing energy conversion efficiency, a well-engineered system manages critical concerns, such as fuel crossover, mass transport limitations, and electrode poisoning, ensuring stable and trustworthy performance over extended periods of time. However, efficiency is the main feature that draws people to DLFCs because it directly affects their economic and environmental viability. Direct light fuel cells (DLFCs) with high efficiency can achieve

DOI: 10.1201/9781003512592-2

higher power outputs while using less fuel and emitting fewer emissions. This is achieved by optimizing energy conversion procedures and managing losses. As a result, DLFCs are now more competitively priced than conventional power production technologies. By focusing on system design and efficiency, researchers and engineers can thereby accelerate the development and acceptance of DLFCs and usher in a new era of clean, sustainable energy production and utilization (Figure 2.1) [2,3].

The junction of efficiency, sustainability, and system design becomes critical in the quest for a sustainable energy future, especially when it comes to DLFCs. This chapter explores the complex interplay between design principles, environmental considerations, and operational effectiveness that is necessary to maximize these advanced energy systems. DLFCs, which use liquid fuels to generate electricity with less of an impact on the environment, are a promising new direction in clean energy generation. But realizing their full potential will need a thorough comprehension of the complex design decisions that influence how well they function, the sustainability implications of their use, and the pursuit of higher energy conversion efficiency.

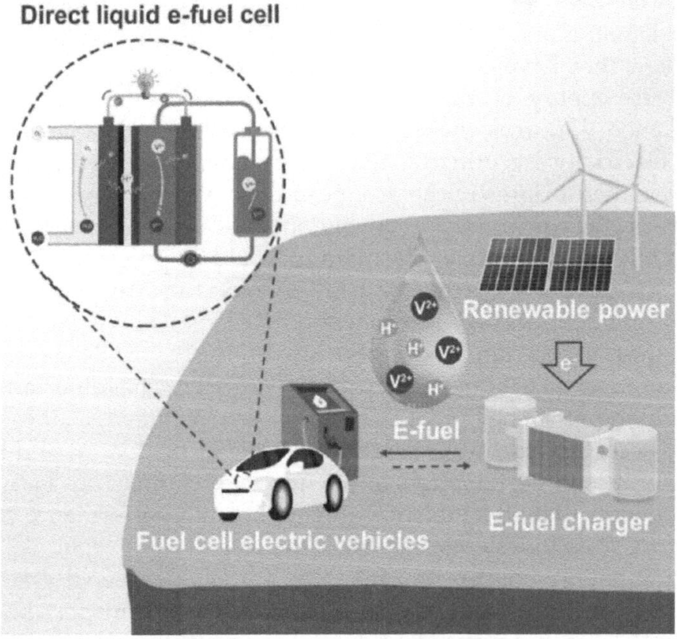

FIGURE 2.1
Recent progress in fuel cell technology, DLFC, oxide.

Source: (*Reproduced with permission. (Shi et al., 2020) [30] Copyright 2024)

By starting this investigation, we can reveal the complex workings of direct liquid fuel cell technology and open the door to a more effective and sustainable energy environment.

2.2 System Design

2.2.1 Importance of System Design on Performance of DLFCs

System design is crucial for DLFCs because it helps maximize efficiency, guarantee compatibility with liquid fuels, improve energy density, and meet the unique requirements of small-scale and portable applications. A thoughtful system design can enhance the overall efficacy and usefulness of direct light fuel cells. DLFC design is a very important field that affects many important factors that affect how well these energy systems work and how long they last. By coordinating component integration and optimizing the placement of essential components, a well-optimized system design plays a critical role in enhancing the overall efficiency of DLFCs. To address the growing need for sustainable and energy-efficient power solutions, this efficiency is essential. The compactness and portability of DLFCs are influenced by the system architecture, which also facilitates the smooth integration of various liquid fuels, including methanol, ethanol, and formic acid, providing flexibility for a wide range of applications. Careful system design has led to this compact form factor, which is essential for applications ranging from small-scale power systems to portable electronic gadgets. Furthermore, improvements in system design solve issues related to thermal management, enabling DLFCs to function at room temperature. This simplifies the devices' construction and operation while improving security and dependability.

2.2.2 System Design of Direct Liquid Fuel Cells

To enable the electrochemical reaction between the liquid fuel and the oxidant, which produces electrical energy, DLFCs are made up of a few crucial parts. Depending on the type of DLFC, the components may differ, but these are the general elements of a typical DLFC (Figure 2.2):

2.2.2.1 Fuel Cell Stack

This is the central element in which electrochemical reactions occur. It is made up of several separate cells joined together in series. An anode, a cathode, and an electrolyte are present in every cell (Figure 2.3).

The liquid fuel undergoes electrochemical oxidation at the anode. Hydrocarbons, such as methanol and ethanol, are common forms of liquid

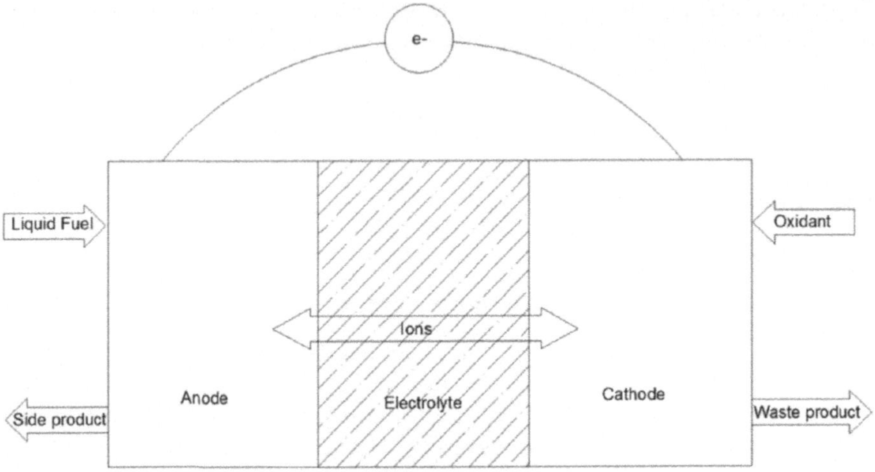

FIGURE 2.2
General working principle of DLFCs.

Source: (*Reproduced with permission. (Ong et al., 2017) Copyright 2024)

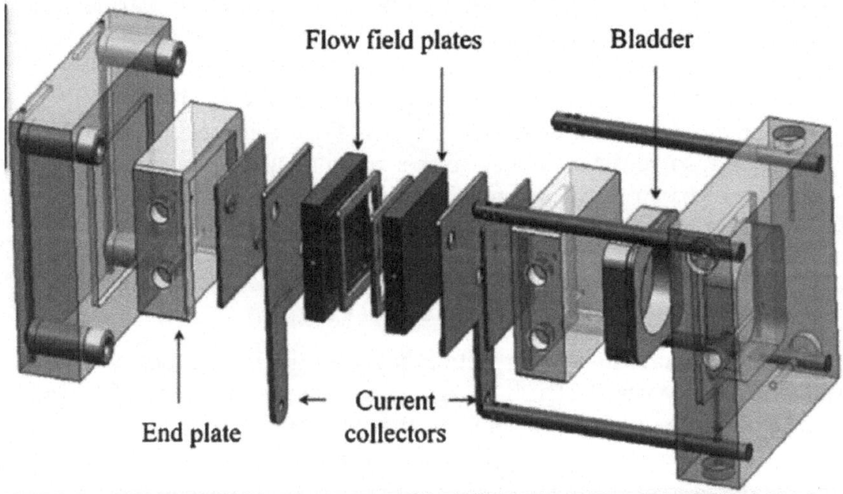

FIGURE 2.3
General cell stack of DLFCs.

Source: (*Reproduced with permission. (Ong et al., 2017) Copyright 2024)

fuels. The fuel's chemical bonds are broken more easily by the anode catalyst, releasing protons, electrons, CO_2, and other byproducts.

$$C_2H_5OH \text{ (ethanol)} + 3H_2O \rightarrow 6H^+ + 6e- + 2CO_2$$

An oxidizing agent (typically airborne oxygen) is reduced at the cathode. Water is usually produced as a byproduct of the reaction between oxygen and electrons, which is aided by the cathode catalyst.

$$O_2 + 4e- + 4H^+ \rightarrow 2H_2O$$

An electrolyte or membrane is used in direct liquid fuel cells to help move ions from the anode to the cathode. Proton exchange membranes (PEM) and solid oxide materials are examples of common materials. A system is needed for DLFCs in order to supply liquid fuel to the anode. This could entail a pump or some other kind of mechanism to guarantee a steady and regulated fuel supply. For the electrochemical reaction to occur at the cathode, an oxidant typically air or oxygen is required. To guarantee a constant supply of air, this may entail the use of a fan or another device.

Bipolar plates allow electrons to flow between fuel cell stack cells individually. Additionally, they provide fuel and oxidant, or reactants, to each individual cell.

Current collector materials act as conductors, gathering and directing the electrical current produced within the fuel cell towards an external circuit.

Gas diffusion layer is composed of porous materials that support good electrical conductivity while allowing reactants (fuel and oxidant) to flow to the appropriate electrodes. DLFCs frequently include a cooling mechanism to control the heat produced during the electrochemical processes and keep the system running at its ideal temperature.

Electrical energy is produced by the combined electrochemical reactions at the anode and cathode. An electric current is produced as the electrons move through an external circuit.

$$C_2H_5OH + O_2 \rightarrow 2CO_2 + 3H_2O$$

To regulate the fuel and oxidant flow rate, temperature, and pressure, the system may consist of parts like pumps, fans, and controls. The fuel cell's performance is enhanced by these controls. Fuel cell efficiency and energy density are optimized by designers, while system complexity, cost, and dependability are considered.

2.2.3 Working Principle of Direct Ethanol Fuel Cell

Ethanol is directly supplied to the anode chamber of DEFC systems in liquid or vaporized form, regardless of its anhydrous or diluted state. A direct feed arrangement is produced simultaneously by introducing air into the cathode chamber.

Ethanol is oxidized at the electrode/electrolyte interface, producing protons, carbon dioxide, and electrons, as well as partially oxidized ethanol byproducts, like acetaldehyde and acetic acid. The kind of electrolyte used determines the particular reactions; proton-conducting electrolytes are a popular option.

The oxidation of ethanol results in the production of carbon dioxide, protons, and electrons during the anode/electrolyte interface reaction. The following equation represents the entire oxidation process:

$$C_2H_5OH + 3H_2O \rightarrow 2CO_2 + 12H^+ + 12e^-$$

At the cathode/electrolyte interface, protons, electrons, and oxygen from the air combine to form water. This reaction is expressed as follows:

$$3O_2 + 12H^+ + 12e^- \rightarrow 6H_2O$$

2.2.3.1 Overall Reaction

Combining the anode and cathode reactions, the overall electrochemical process in a DEFC can be summarized as follows:

$$C_2H_5OH + 3O_2 \rightarrow 2CO_2 + 3H_2O$$

Since ethanol participates directly in the electrochemical reaction, the kinetics of ethanol oxidation in DEFC systems can be slow because oxidation requires the transfer of 12 electrons. The reaction could, however, continue to intermediate phases, like acetaldehyde or acetic acid, which require fewer electrons (2 or 4, respectively), and this could have an effect on the DEFC's overall performance and efficiency [4,5].

2.2.4 Types of System Design

2.2.4.1 Active System

Active systems include additional parts (balance-of-plant) like a fan and a pump. These parts are used in the fuel cell to provide functions like humidification, cooling, and product and reactant control. Active systems can function in conditions with favorable temperature, pressure, concentration, and flow rate when they have additional components included. Electrochemical activity and fluid mass movement are enhanced by active systems. This improvement causes the direct liquid fuel cell to produce more power and have a higher current density. There are compromises with active systems even with their performance advantages. One trade-off is reduced system energy density combined with increased system complexity. Larger fuel cells are typically thought to be a better fit for active systems, indicating that in larger-scale applications, the benefits might exceed the disadvantages [6].

2.2.4.2 Passive System

Diffusion, evaporation, convection (air breathing), and natural capillary forces are the foundations of passive systems. These systems complete all tasks without requiring extra electricity. Generally, low current density operation of passive systems leads to lower cooling load, less problems with water management, less heat production, and lower necessary fuel delivery rates. In passive systems, a well-thought-out compact architecture can lead to maximum system energy density, reduced costs, increased fuel utilization, and higher reliability. Because passive systems have advantages in terms of cost, fuel efficiency, energy density, and reliability, they are thought to be more appropriate for portable power sources [6].

2.2.5 System Design and Fabrication

Direct ethanol fuel cell (DEFC) hardware components after being separated. DEFC components consist of endplates, current collectors, flow field plates, and bladders. In between the flow field plates, there are membrane electrode assemblies. An alkali anion-exchange membrane (AAEM) or proton exchange membrane (PEM) stacked with flat plate electrodes and a catalyst is called a membrane electrode assembly (MEA). The anode diffusion layer, anode catalyst layer, electrolyte membrane, cathode catalyst layer, and cathode diffusion layer are the five theoretical components of the MEA [7]. Diffusion layer: provides the electrodes with their fundamental mechanical structure and acts as a conduit for reactants and products. It serves as both a conductor and a current collector. Catalyst layer: the layer where electricity is produced by electrochemical processes. It permits the passage of electrons and protons while offering an active surface and acting as a conduit for reactants and products. Electrolyte membrane: between the anode and cathode compartments, the electrolyte membrane serves as a separator and a transporter of protons. Common membrane materials, like Nafion, can have their performance improved by adding KOH or NaOH. The preparation of MEAs depends on elements like anode liquid solution, anode electrode structure, and electrode area. Fuel cell performance can be greatly impacted by preparation methods, and problems, such as electrode delamination, can occur. Research on MEA preparation techniques, like brushing and decal transfer, reveals important impacts on ethanol crossover and DEFC performance. Research on various membrane materials and non-platinum electrocatalysts reveals a range of characteristics and performances. Currently, two types of membranes are used: PEM and anion-exchange membranes (AEM). While PEM, which typically uses Nafion, has been widely used, AEM has issues with stability and durability [1,7].

2.2.5.1 MEA Stability Enhancement

The addition of inorganic phases to polymeric networks has been investigated by researchers to improve membrane stability and reduce swelling characteristics. Neodymium triflate, for example, is a modification used in Nafion composite membranes to lower fuel permeability without sacrificing proton conductivity. AEM DEFC is renowned for its high-power density; however, cathode flooding and durability are issues. A nickel foam layer has been added recently to improve ion conductivity and performance, decrease diffusion resistance, and improve structure. The MEA's components and preparation techniques have an impact on the fuel cell's overall performance and efficiency, making it a crucial component of the DMFC system. In order to overcome obstacles and improve the stability and performance of direct liquid fuel cells, research is still being done on optimizing MEA design and materials [1,7,8].

2.2.5.2 Anode Catalyst

2.2.5.2.1 Palladium-Based Catalyst

Pd serves as the primary electrode in more recent anode catalysts for acid media DEFC. PdCoMo/C, porous Pd particles (nanoPd), and Pd/Nickel foam (3D) are a few examples of Pd-based catalysts. Pd-based catalysts' high stability and large surface area make them promise for enhancing ethanol electro-oxidation. PdPt/Ru, Pd-MWCNT-Ni, $PdMn_3O_4/C$, heterobimetallic Ru/Pd complexes, $PdCeO_2$, $PdZrO_2$, Pd_3Au/C, and other Pd-containing catalysts are among the others.

2.2.5.2.2 Alkaline Media DEFC

Compared to acid media, alkaline media DEFCs have the advantage of enabling the use of catalysts that are not Pt-based, which could lower costs and increase efficiency. Pd has been shown to be useful in the electro-oxidation of ethanol in alkaline media; other metal adatoms being investigated include Pb, Tl, Cd, Ru, Au, Sn, and Cu. $PdCeO_2$, $PdZrO_2$, PdTb/C, Pd_3Au/C, PdMgO, $PdCo_3O_4$, $PdMn_3O_4$, PdNiO, $PdIn_2O_3$, PdBi/C, and PdAg/C are a few examples of catalysts in alkaline media.

The degradation of anode catalysts in alkaline media poses a concern, as it affects the overall performance and electrochemical active surface area. It is proposed that because OH^- ions replace carbonate ions in alkaline media DEFC, CO_2 formation will be higher than in acidic media [1,7].

2.2.5.2.3 Platinum-Based Catalyst

Platinum (Pt) is primarily used in acidic media; however, because of its sensitivity to CO adsorbed species poisoning, its use has become less common. Ethanol electro-oxidation can be improved by modifications like utilizing catalysts, like $PtRuO_2/C$, or integrating with other species, like Ru, Sn, Pd, and W. PtRuMo/C, PtReSn/C, PtRuNi/C, PtSnNi/C, $PtSTiO_2/MWCNT$, and $PtSSnO_2/MWCNT$ are some of the more recent Pt-related catalysts.

2.2.5.3 Other Catalysts

The function of Sn in CO_2 selectivity, performance, and ethanol electro-oxidation are significantly improved by Ir_3Sn/C and PtSnIr/C. As catalysts for breaking CeC bonds, rhenium (Re) and rhodium (Rh) are investigated; Re appears to have potential for enhancing activity in acidic media (Figure 2.4) [1,7].

2.2.5.3.1 Cathode Catalyst

$AgeW_2C$, Pd, and PteRu are the best cathode catalysts, according to Antolini and Gonzalez, for direct ethanol fuel cells (DEFCs). Catalysts that perform better than Pt or Pt/C are PteCo/C and PtePd/C. PteCo/C (E-TEK) exhibits good performance at 60–100°C, while PtePd/C outperforms Pt at 90°C. In DEFCs, PteCo/C (E-TEK) outperforms Pt/C in terms of cathode catalyst performance when used at 60–100°C. When used as a cathode material in DEFCs, PtePd/C performs better than Pt at 90°C. The creation of bifunctional structures uses highly dispersed gold layers deposited as cathode catalysts on layers of nickel-chromium

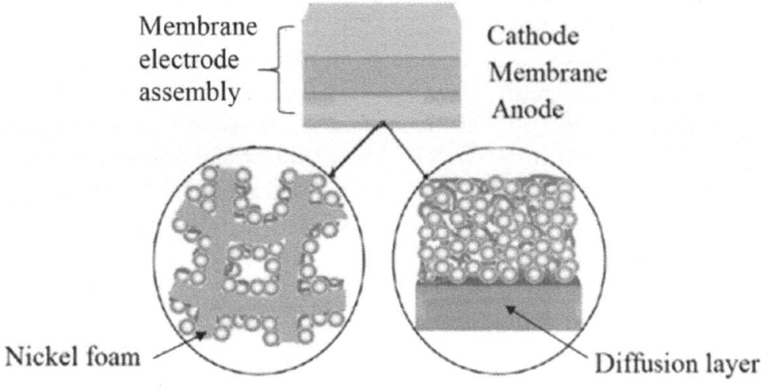

FIGURE 2.4
Schematic diagram of left-conventional anode structure and right-nickel foam structure at anode layer.

Source: (*Reproduced with permission. (Tan et al., 2018) [9] Copyright 2024, 2024)

(Ni-Cr) foam. In a study showcasing these structures' potential for direct ethanol fuel cells, An et al. achieved power densities of 200 mW cm^{-2}, a 48% increase over conventional cathode catalysts (135 mW cm^{-2}) [1,7].

2.3 Sustainability of DLFC

Alternative, renewable energy sources, like sunshine, wind, and water, are essential due to the limited availability of traditional energy resources, like uranium and fossil fuels, as well as the negative environmental effects associated with them. Furthermore, wastes and biomass fuels with an emphasis on waste-to-energy technologies are investigated as possible sustainable energy sources [10,11]. When looking for sustainable energy options, DLFCs have shown promise. For DLFCs to remain viable as clean energy sources over the long term, sustainability is essential.

2.3.1 Environmental Impacts

When it comes to the environment, DLFCs are superior to conventional energy sources. Because DLFCs generate energy with fewer greenhouse gas emissions than combustion-based systems, they help to mitigate climate change and reduce air pollution. The use of liquid fuels, like ethanol or methanol, reduces the number of pollutants released while using traditional fossil fuels. These days, research delves into DLFC life cycle analysis, offering insights into their whole environmental impact [12,13]. By concentrating on the materials utilized in the building of DLFCs, the excerpt draws attention to a critical component of improving DLFC sustainability. Researchers are working hard to find ways to make DLFC components that are recyclable and sustainable. These days, research focuses on creating DLFC components using environmentally friendly materials as a primary consideration. The main goals are to solve environmental issues and encourage prudent resource use when building DLFCs. The goal of the project is to reduce the environmental impact of producing and disposing of DLFC components by using recyclable and sustainable materials. This strategy emphasizes the significance of considering the full life cycle of DLFC technology, which is in line with more general sustainability aims [14,15]. The studies essentially add to the long-term sustainability of DLFCs by incorporating a material perspective that gives preference to environmentally beneficial decisions. This is essential to guarantee that, throughout its lifecycle, DLFC technology not only offers effective energy solutions but also does so in an environmentally conscious and sustainable way. It provides guidance for DLFC technology advancements in the future, pushing the sector towards more environmentally friendly operations [15,16].

2.3.2 Life Cycle Analysis

Life cycle analysis is used in a thorough sustainability study of direct liquid fuel cars, considering the environmental impact at each stage of the process, from raw material extraction to production, operation, and disposal. Examining every stage of the life cycle is how researchers hope to improve the overall performance of DLFCs in terms of efficiency and environmental impact. This strategy aids in the creation of more environmentally friendly and long-lasting energy solutions [13]. The type of fuels that DLFCs use has a significant impact on how sustainable they are. The overall environmental profile of DLFCs is improved by integration with renewable fuels, such as biofuels or synthetic fuels made from sustainable sources. This strategy fits with the overarching objective of lowering dependency on non-renewable resources and moving towards a circular economy. Plans for recycling and appropriate disposal of DLFCs at the end of their useful lives are also necessary to ensure their sustainability. Technologies that allow for the recovery and reuse of important components are being developed in an effort to reduce waste and minimize the impact on the environment.

2.3.3 Sustainable Design

The materials utilized in the building of DLFCs have a major impact on their sustainability. The goal of ongoing research is to provide abundant, affordable, and ecologically friendly substitute materials. The investigation of naturally occurring catalysts, environmentally acceptable membrane materials, and recyclable parts helps to lessen the environmental effect of DLFC operation and production [14,16,17]. Sustainability and energy efficiency are linked in DLFCs. Optimizing system design, electrode configurations, and operational parameters is part of the effort to achieve improved efficiency. Enhanced efficiency raises the long-term sustainability and economic viability of DLFCs by lowering resource consumption and improving overall performance. DLFCs must be scalable and easily incorporated into the current energy infrastructure to be genuinely sustainable. Scalability guarantees that DLFC technology can satisfy the expanding energy requirements of various applications, ranging from industrial systems to portable gadgets. Compatibility with current energy storage systems and grids is a necessary component of integration, which makes the shift to more environmentally friendly energy practices easier.

2.3.4 Technological Advancement

Current research emphasizes how crucial it is to keep developing DLFC technology to maximize effectiveness and reduce environmental effect. Recent developments in electrode materials, system optimization, and DLFC design demonstrate advances that enhance performance while also adding to the

DLFCs' overall sustainability. The emphasis on environmentally friendly electrode materials and system optimization suggests a conscious attempt to balance technical advancement with environmental stewardship. This acknowledgment of sustainability as an essential factor in the development of DLFCs marks a change in the direction of more accountable and effective energy solutions, which will impact the course of upcoming advancements in the industry [11,16]. Over a four-year operating span, a study compares the viability of using direct methanol fuel cells (DMFCs) vs. lithium-ion batteries to power portable electronics. Li-ion batteries are less expensive at first, but over time, the DMFC system turns out to be more economical. Although the weights of the two systems are identical, the DMFC needs less room to store more energy. The study emphasizes how the DMFC has a better energetic efficiency, which lowers CO_2 emissions and improves sustainability overall. DMFC efficiency is further increased by improved specific power and reduced fuel crossover, making it a more advantageous long-term solution than the battery system. The feasibility research, which emphasizes cost reductions, comparable weight, and enhanced sustainability over an extended operating term, promotes the deployment of DMFCs for portable device power overall [18,19].

Government policies and regulatory frameworks that encourage the development and uptake of renewable energy technologies have a major impact on sustainability in DLFCs. In order to promote sustainability, supportive policies are essential. Researchers and policymakers work together to establish an atmosphere that supports the development of sustainable energy solutions, such as DLFCs. Furthermore, public acceptability and awareness are a necessary component of the inextricably related feature of DLFC sustainability. Promoting acceptance requires educating the public about the long-term sustainability, technological advances, and environmental advantages of DLFCs. Support from the general public not only encourages investment and innovation but also speeds up the shift to a more sustainable energy environment, thereby solidifying DLFCs' place as a major participant in the clean energy paradigm [18,14,20].

Furthermore, the UN's Sustainable Development Goals (SDGs) highlight urgent global concerns, and DLFCs' sustainability provide a wonderful opportunity to address them. DLFCs provide a path towards achieving several SDGs by providing a clean and efficient energy solution that reduces reliance on fossil fuels, reduces greenhouse gas emissions, and promotes environmental stewardship. DLFCs use a variety of liquid fuels, including renewable and bio-derived sources, to assist meet Goal 7 (Affordable and Clean Energy) by diversifying the energy mix and expanding access to clean power, particularly in off-grid and isolated locations. Furthermore, their high efficiency and low emissions are in line with Goal 13 (Climate Action), which helps in the battle against climate change and its detrimental impacts (Figure 2.5). They offer a sustainable alternative to combustion-based technology. Additionally, by promoting infrastructure development

FIGURE 2.5
United Nations Sustainable Development Goals (reused from an open-access journal).

and technological innovation, DLFCs can be used for a variety of purposes, including portable electronics and stationary power generation. Goal 9 (Industry, Innovation, and Infrastructure) is furthered by this. DLFCs are a tangible step towards achieving several SDGs since they offer a sustainable energy solution that not only addresses pressing environmental challenges but also promotes social progress and global economic growth.

To sum up, DLFC technology sustainability requires an all-encompassing strategy that considers materials, energy efficiency, economic feasibility, scalability, regulatory backing, and public awareness. As the area develops, the incorporation of these components will make DLFCs a key component of the sustainable energy paradigm, helping to create a future that is greener, more efficient, and ecologically conscious.

2.3.5 Efficiency of Direct Liquid Fuel Cells

Fuel cells are composed of an anode for fuel intake, a cathode with an oxidant (often oxygen), and an electrolyte to facilitate ion movement between the anode and cathode. Fuel cells are anticipated to be a viable energy conversion technology for the next century. Fuel cell efficiency, in contrast to that of combustion engines, is dictated by the Gibbs function change to the enthalpy change ratio in the total cell reaction rather than the Carnot efficiency limit. The efficiency of an internal combustion engine is significantly lower than that of a hydrogen/oxygen fuel cell. Voltage efficiency, which is the ratio of the actual voltage under operating conditions to the theoretical cell voltage, is a commonly used practical metric to quantify fuel cell efficiencies. There are several distinct kinds of fuel cells that use different electrolytes and fuels; fuels high in hydrogen are most popular. Although solid oxide ceramic fuel cells, phosphoric acid fuel cells, and proton exchange membrane fuel cells have found commercial use, issues with CO poisoning, water management, and the need for noble catalysts still exist.

To overcome these obstacles, work is being done, especially in the area of creating novel fuel cell materials that are affordable, easily obtainable, CO-tolerant, and extremely efficient. Direct liquid fuel cells have garnered a lot of attention lately due to their noteworthy benefits, which include low operating temperatures, great energy conversion efficiency, and little environmental impact. Direct methanol fuel cells, direct ethanol fuel cells, and direct formic acid fuel cells are some of the subcategories of these fuel cells that are based on the kind of fuels that are provided to the anode [3,6,11].

2.3.6 Factors Influencing Efficiency of Direct Liquid Fuel Cells

Attempts to improve the direct liquid fuel cell's efficiency have included the methanol oxidation reaction's catalyst, such as Pt. Pt's effectiveness is maximized through the construction of size-controlled nanostructures with precise distribution patterns, despite its high cost and vulnerability to CO poisoning. Because of their tendency to aggregate, surfactants or polymer holders must be added in order to attain a uniform dispersion—albeit at the expense of higher internal resistance [11,21]. Pt-coated carbon nanotubes (CNTs) have been made by an atomic layer deposition technique to overcome these challenges. The resultant catalyst showed a peak current density of 2000 mA/mg Pt. Its best properties were obtained by using deposition circles and annealing conditions. With excellent results for the ethanol oxidation reaction, similar procedures were used for Pd nanoparticles stabilized by multi-walled carbon nanotubes (MWNTCOOH) without surfactant. Pd-based catalysts have demonstrated potential in catalytic activity, outperforming Pt [21].

Bimetallic nanoparticles supported by carbon nanotubes, as PtPd and PdNi, are the subject of future developments. These catalysts demonstrate the potential of alloy or composite nano catalysts by performing better for the oxidation of ethanol. The investigation includes creating PdO/MWCNTs with extremely high electrocatalytic activity that outperforms Pt/C catalysts available on the market in terms of forward peak current density. PdO/MWCNT inclusion also exhibits catalytic activity for formic acid and methanol oxidation processes. FePd nanoalloys on g-Fe_2O_3, anchored on MWCNTsCOOH, have been synthesized because of extensive research on the ethanol oxidation reaction. Peak current density increased significantly as a result of this method's improved poisoning tolerance and decreased charge transfer resistance [21].

Pd-decorated MWCNTs were used for direct formic acid fuel cells, and the electrocatalytic activity was maximized by balancing the absorption of sulfuric and formic acids. MWCNTs with bimetallic PdNi nanoparticle decorations performed better as well, highlighting the significance of catalyst composition [21,11]. The potential of gold as a catalyst in fuel cells and its relationship to the pyrolysis duration during preparation are demonstrated by the electrocatalytic capabilities of Au/MWCNTs for ethanol oxidation in alkaline solution and oxygen reduction in neutral solution. Nanostructured materials have a significant impact on electrode performance, highlighting their role in obtaining ideal porosity, high surface area, and superior electrical conductivity. These qualities are essential for guaranteeing increased reactant flow, stability, and general performance in the demanding fuel cell environment [21].

Electrocatalysis has undergone a revolutionary change thanks to nanostructured materials, which are distinguished by their distinct characteristics and nanoscale dimensions. Regarding DLFCs, these materials offer several significant benefits: *High surface area:* The surface area per unit volume of nanostructured materials is remarkably high. The effectiveness of catalytic processes at the electrode is improved by the increased surface area, which offers more active sites for electrochemical reactions. *Good electrical conductivity:* During electrochemical reactions, effective electron transmission is ensured by the intrinsic conductivity of nanostructured materials, which are frequently produced from carbon-based structures, like nanotubes or graphene. As a result, the cell performs better overall and has less electrical resistance. *Appropriate porosity:* The porosity of nanostructured materials is regulated, facilitating the effective movement of reactants to the catalytic sites. This characteristic plays a pivotal role in preserving a high flux of reactants, expediting reaction rates, and augmenting cell efficiency. The use of nanotechnology in catalyst design signifies a substantial advancement in DLFC technology. The use of nanomaterials has greatly contributed to recent advancements in catalyst technology. By modifying nanostructures to increase catalytic activity and selectivity, researchers have addressed major issues with fuel cell performance [4,18,11]. The incorporation of nanostructured materials into DLFC electrodes is a big step in the right direction for increased performance and efficiency. More developments in electrode engineering and catalyst design are expected as this fascinating field of study progresses, opening the door to more efficient and environmentally friendly direct liquid fuel cells [18,17].

2.3.7 Heat Management

Direct liquid fuel cells work by using electrochemical processes to transform the chemical energy of liquid fuels into electrical energy. Optimizing heat management is essential to extending the life and performance of direct light fluorescent lights. Heat is produced by DLFCs by nature when electrochemical reactions take place at the anode and cathode. While some heat is necessary to maintain ideal operating temperatures, too much heat can cause material damage, decreased efficiency, and performance degradation [22, 23, 11]. Because fuel oxidation and oxygen reduction are exothermic processes, DLFCs have thermal problems. Controlling heat is essential to avoiding thermal runaway, which can shorten cell life and cause system failure [22,24,11]. Several techniques have been developed to efficiently control heat in DLFCs. Cooling systems: Including efficient cooling mechanisms, like air or liquid cooling, aids in removing extra heat from the fuel cell. Robust thermal control mechanisms guarantee consistent temperature dispersion throughout the cell. *Heat exchangers:* Using heat exchangers enables heat to be transferred between the fuel cell system's various components. This helps to improve overall energy efficiency by keeping electrochemical reactions operating at ideal temperatures and repurposing excess heat. *Advanced materials:* For effective heat dissipation, high thermal conductivity materials must be chosen. Improved heat management is facilitated by novel materials, such as thermal interface materials and sophisticated ceramics [25].

2.3.8 Integration of Smart Controls

Using smart control systems to optimize heat management in DLFCs is becoming more and more popular. These systems provide effective heat dissipation without sacrificing performance by dynamically adjusting operational settings in response to current circumstances. The effectiveness of DLFCs depends critically on effective heat management. Current investigations and technological developments persist in tackling heat issues and enhancing the overall effectiveness of these fuel cells. It is anticipated that as the area develops, advancements in cooling systems, smart controls, and materials will be crucial in improving heat management and expanding the use of DLFC technology in real-world applications [22].

2.4 Challenges and Future Progress of DLFC Technology

DLFCs have potential to replace conventional energy sources, but commercialization is hindered by limitations. To position DLFCs as renewable energy alternatives, comprehensive solutions must address specific challenges. The industrial application landscape faces competition with batteries and fossil fuels. This section acknowledges the evolving potential of DLFCs and the importance of addressing these challenges, concluding with a proposal that serves as a guide for future researchers in overcoming obstacles and realizing the full potential of DLFCs [19]. DLFCs face challenges due to high cost and dependency on noble metal-based catalysts, particularly platinum, for optimal performance. Commercial applications use PtRu and Pt catalysts, generating 23 mW cm^{-2}. Reducing catalyst loading can achieve power densities of 14 mW cm^{-2} and 12 mW cm^{-2}, respectively. Reducing loading in DLFCs negatively impacts system performance, leading to high-loading catalysts and increased system costs. Researchers are exploring catalyst modifications using carbon nanofibers, mesoporous carbon, and graphene to improve efficiency. However, ongoing research is crucial as catalyst-related issues remain a major hurdle for DLFC commercialization, directly impacting system costs. Fuel crossover is a major problem for DLFCs, especially in liquid fuel feed systems, where fuel molecules move from the anode to the cathode due to the natural concentration gradient. This event results in a decrease in total fuel consumption, which negatively impacts cathode activity, produces excess heat, and reduces system performance overall. Operating factors, like temperature, pressure, and flow rate, affect how severe fuel crossover is; hence, maintaining DLFC efficiency depends on optimizing fuel concentration. Methanol fuel crossover is a common problem in DMFC systems that causes a 30% decrease in efficiency. To address this issue, expensive upgrades to membranes and catalysts are required [26].

Researchers are looking into alternative fuel-based alcohols, such as ethanol and propanol, which have lower fuel permeability than methanol and may be able to improve system performance to overcome fuel crossover. Using modified membranes, like sulfonated poly(ether ether ketone) (SPEEK), and substitute materials, like sodium alginate, are two methods. Additionally, efforts are focused on creating effective catalysts that exhibit inert properties at the cathode side; non-platinum catalysts are preferred by researchers. To improve their effectiveness, these substitutes could need more supporting resources. Because of its greater reaction rate, oxidizing agents in particular, hydrogen peroxide are being studied more and more; yet problems, such as decomposition, make it incompatible with some catalysts. Moreover, addressing excess water content and cathode flooding is crucial, and the proposed solution involves the installation of hydrophobic filters in the gas diffusion layer (GDL) to manage water effectively and improve overall DLFC performance [27].

The difficulties and outlook for DLFCs include the crucial issue of chemical safety and environmental worries related to the many kinds of fuels utilized in these systems. A significant safety concern is the toxicity of some fuels. One of the most hazardous fuels is hydrazine, which can seriously damage the kidneys, lungs, and nervous system and can result in fatalities. Because of its high toxicity, hydrazine presents substantial challenges for the commercialization of DLFCs. Certain fuels, particularly those that include carbon and those that have toxic qualities, like methanol and dimethyl ether, raise questions about safety, even when their levels of toxicity are deemed to be within reasonable bounds [28].

In addition, the corrosive qualities of fuels, such as formic acid, give rise to concerns over safety, necessitating cautious handling to avoid direct skin or eye contact. Because some fuels are combustible, there are issues about flammability, which calls for prudence and the creation of fireproof systems for safe use. In addition, hazardous dangers and environmental pollution issues arise from side products produced during fuel reactions, such as ammonia, formaldehyde, glycolic acid, and carbon monoxide. Ongoing efforts are being made to create catalysts that don't interact with harmful byproducts. In the summary, it is emphasized that researchers are growing in favor of ethanol as a relatively safer and more energy-efficient choice and that careful study and evaluation of the possible dangers associated with different liquid fuels in DLFCs are necessary (Figure 2.6).

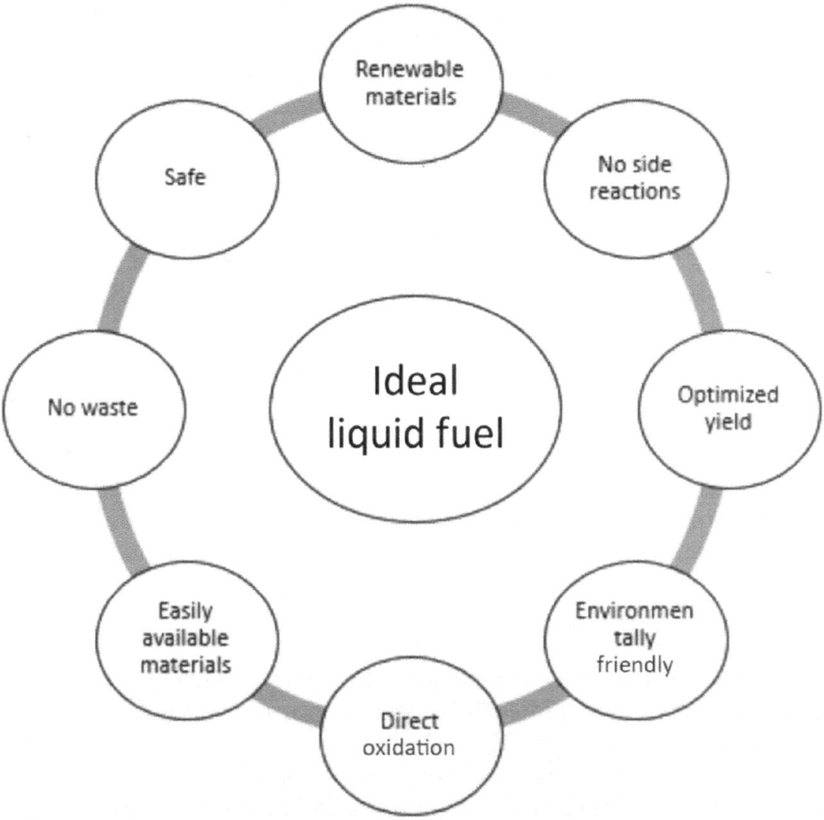

FIGURE 2.6
Characteristics of an ideal fuel.

Source: (Redrawn from open access journal. (Demirci, 2009) Copyright 2024, 2024)

2.4.1 Uncertain Long-Term Durability

A key performance criterion for the successful commercialization of DLFCs is ensuring their long-term durability. Degradation over time is a substantial difficulty to DLFCs, affecting their initial performance. Most DLFC systems use PtRu catalysts in the anode, which can cause problems as the catalyst dissolves and Ru crosses over into the electrolyte, decreasing the cathode's oxygen reduction efficiency. System instability is caused by catalyst agglomeration, electrode porosity variations, and electrolyte membrane degradation. These factors shorten response times and degrade overall performance. Research indicates a decline in performance in several DLFC systems when they are used for longer periods of time. Fuel cell activity is adversely affected when catalyst particle sizes rise and specific surface area falls. System efficiency is further decreased by catalyst deterioration, which also affects oxygen movement, cathode flooding, and water management. It is imperative to address these concerns since extended operation of DLFCs can lead to reduced power density, structural deterioration, and catalyst-related problems. To increase the overall lifespan of DLFC systems, future research should concentrate on stability and durability performance tests. Recent studies have shown encouraging advancements, such as a DMFC system with a lifespan of 20,000 lifetime cycles and enhanced DMFC stack efficiency for longer operation periods. To achieve demonstrated long-term durability, however, is still a significant challenge that needs to be investigated further and optimized for real-world uses [6,19,29].

2.4.2 Production Cost

DMFC market penetration is severely hampered by manufacturing costs, which also make DMFCs less affordable when compared to PEMFCs and Li-ion batteries. Because DMFCs require more expensive metal catalysts (PtRu and Pt) and Nafion-type membranes during production, their production costs are ten times greater than those of comparable Li-ion battery systems. There have been recent initiatives to lower expenses and lengthen the life of DMFC systems. For mobile DMFC devices, the Department of Energy (DOE) has set high standards: 35% system performance, 5,000 hours of life, and \$5–\$7 per watt prices by 2015 [30,6,19].

Even though performance and component optimization have improved, a greater grasp of cost reduction in large-scale industrial production is still needed. The cost of DMFC stacks may be affected by the significant economic advantages of mass production. Reducing the use of precious metals, investigating more affordable membrane alternatives, and improving catalyst and electrode production techniques are some of the cost-cutting strategies to be considered. Fuel cell systems' modular design is emphasized as the best option for integrated mass production. To increase the competitiveness of DMFCs for industrial use and energy applications, the European Community's FP7 DURAMET project has contributed to cost-effective developments, such as techniques for catalyst manufacture and the creation of cost-effective membranes. Achieving optimal competitiveness in large-scale DMFC manufacturing requires careful analysis of all components and factors influencing costs [31].

The main obstacles facing DLFC technology include high production costs, fuel crossover problems, catalyst requirements, safety issues, and uncertain long-term durability. Reaching the full potential of DLFCs as a feasible energy substitute requires overcoming these challenges. Future development depends on cooperative efforts to create affordable production processes, improve catalyst performance, and guarantee system longevity. Notwithstanding obstacles, DLFCs present a viable path towards greener and more sustainable energy options, with the potential to completely change the energy sector through continued research and development [19].

2.5 Conclusion

In conclusion, this book chapter has explored the complex relationship between efficiency, sustainability, and system design in the context of DLFCs. The chapter has emphasized the necessity for comprehensive solutions by addressing issues with catalyst requirements, fuel crossover mitigation, chemical safety concerns, and long-term durability. The successful resolution of these difficulties is closely associated with the search for sustainable energy solutions, wherein developments in manufacturing techniques, catalyst development, and system optimization play crucial roles. In the pursuit of cleaner and more sustainable energy, one promising area of research and development is the effective integration of DLFC technology. The potential for DLFCs to profoundly impact the global energy landscape is becoming more and more evident via continued study, collaboration, and invention. This might help to shape a future where sustainable energy and efficiency are at the forefront of technology breakthroughs.

References

[1] Öztürk, A., Akay, R. G., Erkan, S., & Yurtcan, A. B. (2021). Introduction to fuel cells. In *Direct Liquid Fuel Cells* (pp. 1–47). Elsevier. https://doi.org/10.1016/B978-0-12-818624-4.00001-7

[2] Qian, W., Wilkinson, D. P., Shen, J., Wang, H., & Zhang, J. (2006a). Architecture for portable direct liquid fuel cells. *Journal of Power Sources*, 154(1), 202–213. https://doi.org/10.1016/j.jpowsour.2005.12.019.

[3] Antolini, E., & Gonzalez, E. R. (2010). Alkaline direct alcohol fuel cells. *Journal of Power Sources*, 195(11), 3431–3450. https://doi.org/10.1016/j.jpowsour.2009.11.145.

[4] Badwal, S. P. S., Giddey, S., Kulkarni, A., Goel, J., & Basu, S. (2015). Direct ethanol fuel cells for transport and stationary applications—A comprehensive review. *Applied Energy*, 145, 80–103. https://doi.org/10.1016/j.apenergy.2015.02.002.

[5] Basri, S., & Kamarudin, S. K. (2011). Process system engineering in direct methanol fuel cell. *International Journal of Hydrogen Energy*, 36(10), 6219–6236. https://doi.org/10.1016/j.ijhydene.2011.02.058.

[6] Ong, B. C., Kamarudin, S. K., & Basri, S. (2017). Direct liquid fuel cells: A review. *International Journal of Hydrogen Energy*, 42(15), 10142–10157. https://doi.org/10.1016/j.ijhydene.2017.01.117.

[7] Kamarudin, M. Z. F., Kamarudin, S. K., Masdar, M. S., & Daud, W. R. W. (2013). Review: Direct ethanol fuel cells. *International Journal of Hydrogen Energy*, 38(22), 9438–9453. https://doi.org/10.1016/j.ijhydene.2012.07.059.
[8] Song, S., Wang, G., Zhou, W., Zhao, X., Sun, G., Xin, Q., Kontou, S., & Tsiakaras, P. (2005). The effect of the MEA preparation procedure on both ethanol crossover and DEFC performance. *Journal of Power Sources*, 140(1), 103–110. https://doi.org/10.1016/j.jpowsour.2004.08.011.
[9] Tan, W. C., Saw, L. H., Thiam, H. S., Xuan, J., Cai, Z., & Yew, M. C. (2018). Overview of porous media/metal foam application in fuel cells and solar power systems. *Renewable and Sustainable Energy Reviews*, 96, 181–197. https://doi.org/10.1016/j.rser.2018.07.032.
[10] Hart, D. (2000a). Sustainable energy conversion: Fuel cells—the competitive option? *Journal of Power Sources*, 86(1–2), 23–27. https://doi.org/10.1016/S0378-7753(99)00455-3.
[11] Joghee, P., Malik, J. N., Pylypenko, S., & O'Hayre, R. (2015). A review on direct methanol fuel cells–In the perspective of energy and sustainability. *MRS Energy & Sustainability*, 2(1), 3. https://doi.org/10.1557/mre.2015.4.
[12] Demirci, U. B. (2007a). Direct liquid-feed fuel cells: Thermodynamic and environmental concerns. *Journal of Power Sources*, 169(2), 239–246. https://doi.org/10.1016/j.jpowsour.2007.03.050.
[13] Demirci, U. B. (2009). How green are the chemicals used as liquid fuels in direct liquid-feed fuel cells? *Environment International*, 35(3), 626–631. https://doi.org/10.1016/j.envint.2008.09.007.
[14] Cyril, P. H., & Saravanan, G. (2020). Development of advanced materials for cleaner energy generation through fuel cells. *New Journal of Chemistry*, 44(46), 19977–19995. https://doi.org/10.1039/D0NJ03746J.
[15] Zhang, Y., Anoopkumar, A. N., Aneesh, E. M., Pugazhendhi, A., Binod, P., Kuddus, M., Pandey, A., Kumar Awasthi, M., & Sindhu, R. (2023). Advancements in the energy-efficient brine mining technologies as a new frontier for renewable energy. *Fuel*, 335, 127072. https://doi.org/10.1016/j.fuel.2022.127072.
[16] Hart, D. (2000b). Sustainable energy conversion: Fuel cells—the competitive option? *Journal of Power Sources*, 86(1–2), 23–27. https://doi.org/10.1016/S0378-7753(99)00455-3.
[17] Souza, F. M., Pinheiro, V. S., Gentil, T. C., Lucchetti, L. E. B., Silva, J. C. M., Santos, M. L. M. G., De Oliveira, I., Dourado, W. M. C., Amaral-Labat, G., Okamoto, S., & Santos, M. C. (2022b). Alkaline direct liquid fuel cells: Advances, challenges and perspectives. *Journal of Electroanalytical Chemistry*, 922, 116712. https://doi.org/10.1016/j.jelechem.2022.116712.
[18] Brandon, N. P., Skinner, S., & Steele, B. C. H. (2003). Recent advances in materials for fuel cells. *Annual Review of Materials Research*, 33(1), 183–213. https://doi.org/10.1146/annurev.matsci.33.022802.094122.
[19] Shaari, N., Kamarudin, S. K., Bahru, R., Osman, S. H., & Md Ishak, N. A. I. (2021b). Progress and challenges: Review for direct liquid fuel cell. *International Journal of Energy Research*, 45(5), 6644–6688. https://doi.org/10.1002/er.6353.
[20] Shaari, N., Kamarudin, S. K., Bahru, R., Osman, S. H., & Md Ishak, N. A. I. (2021a). Progress and challenges: Review for direct liquid fuel cell. *International Journal of Energy Research*, 45(5), 6644–6688. https://doi.org/10.1002/er.6353.
[21] Du, H., Zhao, C. X., Lin, J., Guo, J., Wang, B., Hu, Z., Shao, Q., Pan, D., Wujcik, E. K., & Guo, Z. (2018). Carbon nanomaterials in direct liquid fuel cells. *The Chemical Record*, 18(9), 1365–1372. https://doi.org/10.1002/tcr.201800008.

[22] Demirci, U. B. (2007b). Direct liquid-feed fuel cells: Thermodynamic and environmental concerns. *Journal of Power Sources*, 169(2), 239–246. https://doi.org/10.1016/j.jpowsour.2007.03.050.
[23] Wang, H.-N., He, N., Tang, D.-W., & Li, L. (2022). Water and heat management of vapor-fed direct methanol fuel cells with a non-isothermal triple-phase mass transfer model. *Journal of Power Sources*, 552, 232224. https://doi.org/10.1016/j.jpowsour.2022.232224.
[24] Dohle, H. (2002). Heat and power management of a direct-methanol-fuel-cell (DMFC) system. *Journal of Power Sources*, 111(2), 268–282. https://doi.org/10.1016/S0378-7753(02)00339-7.
[25] Rashidi, R., Dincer, I., Naterer, G. F., & Berg, P. (2009). Performance evaluation of direct methanol fuel cells for portable applications. *Journal of Power Sources*, 187(2), 509–516. https://doi.org/10.1016/j.jpowsour.2008.11.044.
[26] Demirbas, A. (2008). Direct use of methanol in fuel cells. *Energy Sources, Part A: Recovery, Utilization, and Environmental Effects*, 30(6), 529–535. https://doi.org/10.1080/15567030600817159.
[27] Jiang, L., Sun, G., Wang, S., Wang, G., Xin, Q., Zhou, Z., & Zhou, B. (2005). Electrode catalysts behavior during direct ethanol fuel cell life-time test. *Electrochemistry Communications*, 7(7), 663–668. https://doi.org/10.1016/j.elecom.2005.04.021.
[28] Li, X., & Faghri, A. (2013). Review and advances of direct methanol fuel cells (DMFCs) part I: Design, fabrication, and testing with high concentration methanol solutions. *Journal of Power Sources*, 226, 223–240.
[29] Qian, W., Wilkinson, D. P., Shen, J., Wang, H., & Zhang, J. (2006). Architecture for portable direct liquid fuel cells. *Journal of Power Sources*, 154(1), 202–213. https://doi.org/10.1016/j.jpowsour.2005.12.019.
[30] Das, V., Padmanaban, S., Venkitusamy, K., Selvamuthukumaran, R., Blaabjerg, F., & Siano, P. (2017). Recent advances and challenges of fuel cell based power system architectures and control—A review. *Renewable and Sustainable Energy Reviews*, 73, 10–18. https://doi.org/10.1016/j.rser.2017.01.148.
[31] Shi, X., Huo, X., Ma, Y., Pan, Z., & An, L. (2020). Energizing fuel cells with an electrically rechargeable liquid fuel. *Cell Reports Physical Science*, 1(7), 100102. https://doi.org/10.1016/j.xcrp.2020.100102.

3

Direct Methanol Fuel Cells; Perspectives in Energy Transition and Sustainability

Ajaz Ahmad Wani, Norazuwana Shaari, Siti Kartom Kamarudin, and Ram K. Gupta

3.1 Introduction

An enormous and crucial task confronting humanity in the twenty-first century is decarbonizing the world's energy systems [1–3]. About two-thirds of the world's carbon dioxide comes from the energy sector, making it an essential sector for combating climate change [4]. At the core of global environmental policy is the goal of reducing carbon emissions since the present period is confronted with the problem of global warming, which is the most prominent ecological crisis [5,6]. Consequently, for a stronger connection and long-term development, it is critical to learn how energy consumption relates to economic growth and to find ways to increase energy efficiency. Since energy is essential for long-term growth, it is in everyone's best interest for countries to prioritize acquiring renewable, low-carbon energy sources. Sustainable energy, also known as energy for sustainable development, is a pressing issue worldwide. A path to renewable energy should be laid out; depending on each country's resources and current situation, it could be different. The Paris Agreement's energy transition goals have made it a reality for all countries. The international community is developing decarbonization strategies to reduce greenhouse gas emissions sustainably [7,8]. The change is influenced by local economic and social factors, making the process distinct for each country. Due to the wide variety of stakeholders and the inherent tensions between their frequently competing interests, the energy transition is intricate and all-encompassing. Natural resource depletion has reached crisis proportions due to the world's increasing energy demands and the widespread use of fossil fuels, which release carbon dioxide into

 DOI: 10.1201/9781003512592-3

the atmosphere. Therefore, building renewable energy sources worldwide is essential for long-term prosperity, particularly in poor nations [9,10].

Fuel cells that use renewable fuels like methanol and hydrogen have garnered much interest as a possible long-term solution. Polymer electrolyte membrane fuel cells (PEMFCs) have attracted much attention in the transportation and portable power industries because they outperform other fuel cell types in terms of power density and ease of cycling [11]. With 30–90% more efficiency (on a "well-to-wheels" basis) than conventional gasoline-powered ICEs, PEMFCs are revolutionizing the transportation industry by producing much more power from a given amount of fuel than conventional ICEs. Most fuel cells used in vehicles are PEMFCs that run on hydrogen. Nevertheless, there are obstacles to the commercialization of hydrogen-powered PEMFC vehicles. These include limitations in onboard storage space (even at 1000 bar, hydrogen takes up approximately 3,000 times more space than gasoline of the same amount) and the absence of a consumer hydrogen distribution infrastructure [12]. DMFCs are the most advanced PEMFC technology for directly converting high-energy-density liquid methanol fuel into electrical power. With the deployment of DMFC, onboard storage limits may be resolved [13]. The current gasoline distribution infrastructure can be used with modest changes to deliver liquid methanol fuel. Despite this benefit, DMFCs are not yet widely used in transportation because of their expensive materials production cost, poor efficiency, and low power density [14]. In contrast to the massive infrastructure expenditures required for a future transportation industry powered by hydrogen, the shift to methanol might be achieved via a succession of smaller stages, a significant advantage of DMFC technology.

Methanol offers much more energy density in volume and gravimetric measurements at a lower cost per unit of energy than liquid hydrogen or compressed hydrogen at 1000 bar [15]. Curiously, 99 grams of hydrogen is present in 1 litre of methanol, which is more than what 1 litre of liquid hydrogen contains (71 g). Gasification and steam reforming processes are the most common ways to produce methanol from various feedstocks, including coal and natural gas. Thermochemical methanol production from biomass resources, including wood, agricultural waste, and other renewable energy sources, has recently gained popularity. These methods yield around 170 gallons of methanol for every metric ton of biomass. Methanol production from biomass is a promising long-term solution due to the renewable nature of the feedstock and the possibility of improving production efficiency. However, compared to natural gas, methanol production from coal and biomass is less efficient, with an efficiency of 50–60% compared to 70–80%. The displacement of crops due to methanol synthesis from biomass is a real worry, however, and it has the potential to escalate into a major problem [16]. After a methanol transportation industry is set up and running, DMFC-powered technology might be a game-changer by outperforming combustion-based, methanol-fuelled vehicles in terms of efficiency, environmental impact, and operational costs [17]. The DMFC is a fascinating, if far-off, transportation power technology because of the appealing qualities of methanol, such as its energy density, simplicity of distribution, and the possibility of generation from clean fossil-based resources (natural gas) or renewable sources (biomass, solar) [18].

The data shown in Figure 3.1 represents the number of research articles published in the last ten years from the Scopus database with the keywords DBFC, DMFC, and direct ethanol fuel cell (DEFC). The rising trend shows that DBFCs are just emerging in the energy sector and are receiving more and more attention from researchers and other sectors. Energy conversion devices employing different liquid fuels, such as borohydride, methanol, and ethanol, are quickly becoming the norm, according to the analysis of the publication record. Because of their usefulness in real-world contexts, this is to be anticipated. Many people are interested in using DLFCs for cars, home appliances, and mobile devices because of their high-power output and the fact that they typically run at temperatures below 80°C. China, the United States, Turkey, and India are the top performers in the bibliometric analyses of nations' publishing research findings on different types of DLFCs; this could be because fuel cell technologies are more widely used in these countries as an alternative energy source than in others (Figure 1a–c). On the other hand, fuel cell technology has become an essential focus in countries like the United States and China due to the rising worldwide need for energy, which is projected to increase by 60% by 2030. Some examples of standard fuel-cell vehicles in China are fuel-cell buses and light haulage trucks. Launching a subsidy on acquiring fuel cell cars under the New Energy Vehicles 2016–2035 program, the Chinese government aimed to promote fuel cell use in China [19]. An increase in fuel cell research in China is anticipated to occur soon due to this strategy. Similarly, fuel cell strategy is booming in the US and other countries throughout various industries, including big and small businesses, from components to systems to integrators to end users. Some governments have used fuel cells to power critical services if the grid shuts down. Several telecom and train businesses maintain the signaling facilities and communication towers. Electric buses powered by fuel cells are already in service in a number of countries. We correctly point out that there has been an excess of fuel cell research, which may explain why fuel cell applications have increased dramatically in developing countries.

Prolific researchers: Figure. 3.1(d–f) highlights the top ten researchers who have conducted extensive studies on methanol, ethanol, and borohydride fuel cells. In the direct methanol and ethanol fuel cells, with 3,881 and 1,424 publications, the most productive researcher is Siti Kartom Kamarudin, who has published 61 and 25 articles, respectively, followed by other researchers. In direct borohydride fuel cells, with 417 publications, Santos, D.M.F. is the first ranked researcher with 31 articles, Šljukić, B. came in second, followed by Chatenet, M. with 23 articles, and in fourth and fifth places, respectively, were Qin, H. and Neto, A.O. with 20 and 19 papers, respectively.

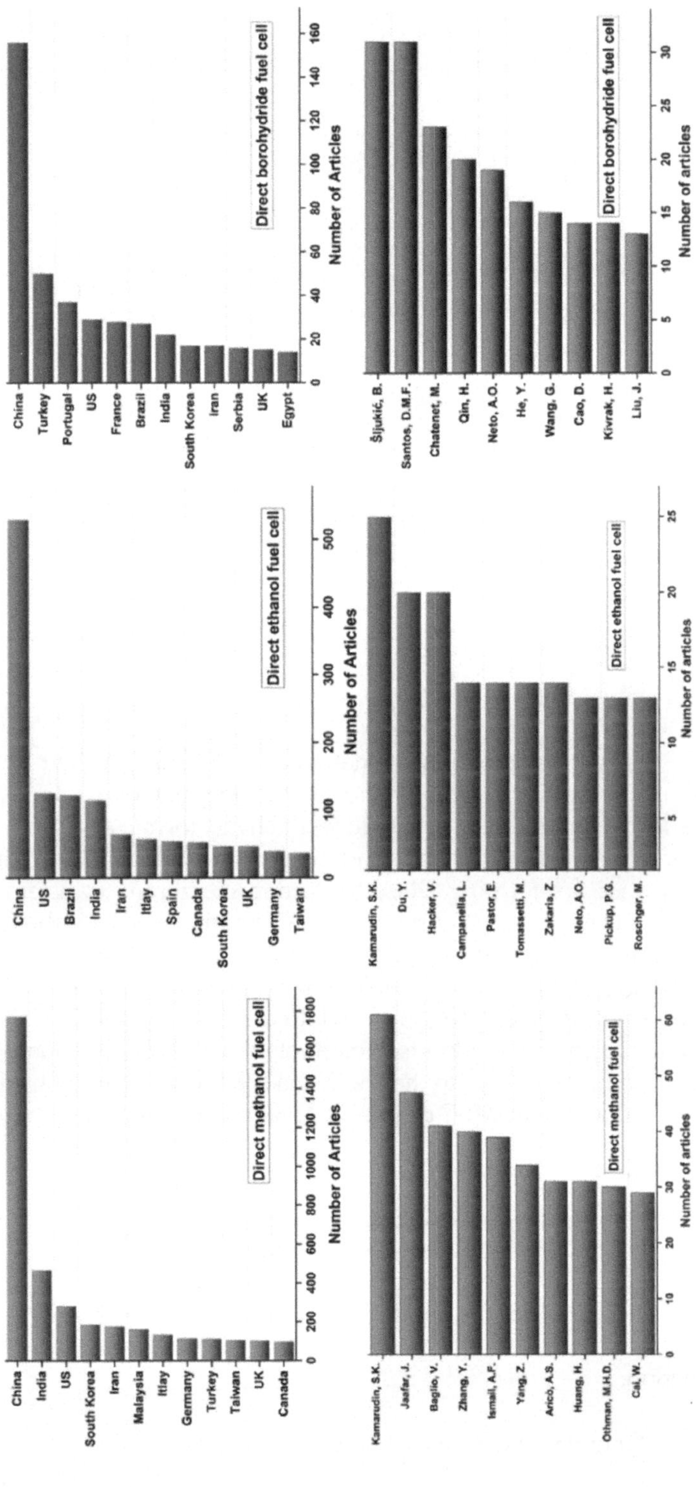

FIGURE 3.1
The progression of DBFC, DMFC, and DEFC publications between 2015 and 2023 (a–c) and top ten researchers in the field of methanol, ethanol, and borohydride fuel cell technology (d–f).

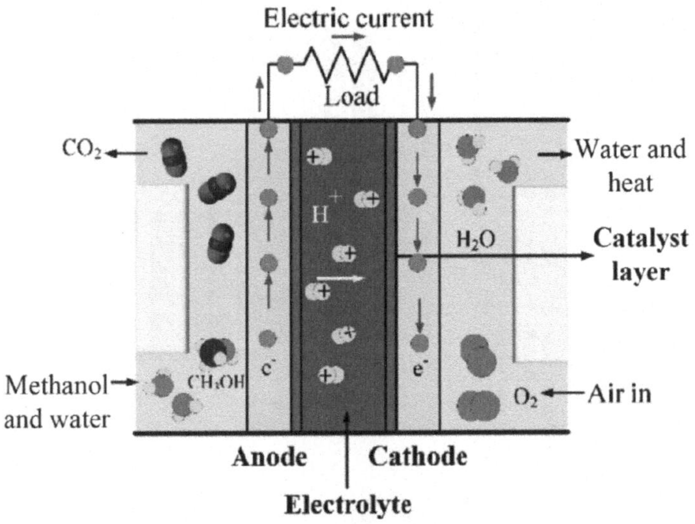

FIGURE 3.2
Schematic representation of direct methanol fuel cell (DMFC). This fuel cell uses methanol as fuel [21].

The DMFC feeds a combination of methanol and water straight to the anode. Direct oxidation of methanol to carbon dioxide may occur by generating intermediary compounds, such as formaldehyde, formic acid, and carbon monoxide. If the anode's methanol oxidation process (MOR) is slow, these intermediate species are forming [20]. The Nafion membrane transfers the protons generated during the MOR process from the anode to the cathode. In contrast, an external circuit transfers the electrons generated from the anode to the cathode (Figure. 3.2). water is created when protons and electrons at the cathode combine with oxygen molecules [21]. In an alkaline DMFC, water is instead generated at the anode due to methanol oxidation by OH^- ions that are transported from the cathode to the anode via the alkaline membrane [22].
The overall reaction in a DMFC can be summarized as follows:

Overall reaction:

$$CH_3OH + \frac{3}{2}O_2 \longrightarrow 2H_2O + CO_2$$

The reactions occurring at the anode and cathode are as follows:

Anode reaction:

$$CH_3OH + H_2O \longrightarrow 6H^+ + 6\,e^- + CO_2$$

Cathode reaction:

$$\frac{3}{2}O_2 + 6H^+ + 6\,e^- \longrightarrow 3H_2O$$

In reality, however, a number of voltage losses cause the cell voltage to fall short of its theoretical values. The slow kinetics of the anode-side methanol oxidation process and the cathode-side oxygen reduction reaction are to blame for the performance losses in DMFC [23]. The total voltage across DMFC cells is much lower than that of PEMFC because methanol crosses from the anode to the cathode, thus lowering the cathode potential. Research and development (R&D) has been devoted to finding active and durable catalysts that minimize kinetic losses, controlling ohmic losses through material selection, engineering, and fabrication, and minimizing mass transport losses through optimal operating conditions in DMFCs [24]. Initial DMFC performance has achieved an acceptable level for practical applications, thanks to ever-increasing knowledge in electrolytes, fuel cell engineering, catalysis, and theoretical understanding of the processes. However, cost and durability are still significant challenges. Verifying a dependable extended lifespan, depending on the application, is crucial for the commercial adoption of DMFC [25].

This chapter is organized to provide a thorough introduction to DMFC technology. It begins with the basic concepts and mechanics of how it works and then moves on to discuss the latest technical developments in the field. Following this, we will go over the efficiency and performance indicators of DMFCs and their environmental and economic advantages, limits, obstacles, prospective applications, market size, and future outlook. An in-depth familiarity with DMFCs and their function in the evolving energy scene is the goal of each part. At the chapter's conclusion, readers will have a firm grasp of the obstacles to the widespread use of DMFCs and a profound respect for the possibilities these fuel cells hold for a more sustainable energy future.

3.2 Fundamentals of Direct Methanol Fuel Cells

Before starting the discussion on different types of fuel cells and their materials, it is necessary to understand fuel cell reactions. Oxygen reduction reactions (ORRs) occur mainly on the cathodes of fuel cells [26]. ORR is the reduction of oxygen to water molecules at the cathode by protons and electrons coming from the anode of the fuel cell. The breaking of the oxygen bond is difficult for ORR due to the high bond energy, which is 498 kJ/mol. Therefore, an efficient catalyst is required to provide an alternate reaction pathway [27]. The reaction involves the formation of intermediates, which can then be converted into water. Depending on the catalyst and working conditions, the ORR mechanism can follow different pathways, which are (a) two-electron transfer pathway in which hydrogen peroxide is formed as an intermediate, (b) four-electron transfer pathway in which oxygen is directly reduced to water, and (c) one-electron transfer pathway where oxygen is first converted to superoxide, which then ultimately gets reduced to water. Hydrogen evolution reaction (HER) is another type of reaction in fuel cells, which occurs mainly at the anode [28]. This reaction involves the oxidation of hydrogen from fuel supplied at the anode. Hydrogen gets oxidized to protons by giving up its electrons, which then travel through the external circuit towards the cathode, while the proton crosses the membrane to reach another side of the cell. The kinetics of HER is almost six times faster than that of ORR. Therefore, nearly ten times more catalyst is required at the fuel cell cathode than at the anode. Another type of reaction in water electrolysis is the oxygen evolution reaction (OER), in which the release of electrons reduces oxygen molecules. These reactions mostly take place during the electrolysis of water [29].

Several essential parts make up DMFCs, and they all work together to keep the cell running smoothly:

1. Anode: Methanol undergoes electro-oxidation at the anode. Complex reactions with several stages produce a wide variety of chemical intermediates. The electro-oxidation of methanol illustrates the reaction pathways and potential intermediates. An essential and performance-limiting stage in a DMFC is the oxidation of methanol. Anodic reaction rate limitation may be achieved with the use of carbon monoxide (CO), an intermediate or byproduct. Platinum (Pt) and ruthenium (Ru) are the most commonly employed catalysts for anode. However, their efficiency is rendered by CO poisoning, the main byproduct of methanol oxidation. As a catalyst for direct metal-hydride fuel cell anodes, platinum is among the most active and widely utilized materials. Researchers are working to develop alternative catalysts with high catalytic efficiency and CO resistance [30].
2. Cathode: Both the ORR and the methanol crossover occur at the cathode side of the fuel cell. A mixing potential caused by methanol in the cathode might reduce overall performance. In a more acidic environment with high temperatures (50–90°C), low pH values (< 1), high potentials (0.6–1.2 V), and operational oxygen partial pressures, DMFC cathodes function. A platinum cathode catalyst is a typical choice, although new research focuses on developing sustainable and economic transition metal-based catalysts [31].
3. Electrolyte: The electrolyte membrane functions as a separator between the anode and cathode, preventing the mixing of methanol flowing from the anode to the cathode. The electrolyte membrane

only allows protons to pass from anode to cathode; therefore, it must be highly proton conductive, impermeable to fuels and oxidants, electrically resistive, and chemically stable. Under long-term operation, the electrolyte should have sufficient mechanical and chemical integrity to avoid developing cracks or pores. The most typical electrolyte membrane for DMFC is Nafion, manufactured by DuPont. Nafion made for DMFCs is classified into three types based on thickness: Nafion 112, 115, and 117 [32].

4. Membrane electrode assembly (MEA): An integral part of a DMFC's overall performance is the MEA. An electrolyte membrane, a gas diffusion layer, and catalysts on the anode and cathode sides comprise the MEA structure (Figure 3.3). The distribution, efficiency, and electrocatalytic activity of the electrocatalysts have direct effects on fuel cell performance, and these factors are significantly influenced by the technique and technology used to prepare the MEA [33]. The catalyst may be applied to the anode and cathode surfaces, such as spraying or scraping [34].
5. Bipolar plates: A bipolar plate is a structure that can collect and distribute current in addition to dispersing reactants. A bipolar plate should withstand acid-base environments, have excellent mechanical strength, and have good conductivity. While metal, graphite, or other composites are the most typical materials for bipolar plates, plastic and ceramic have also been used in some investigations.

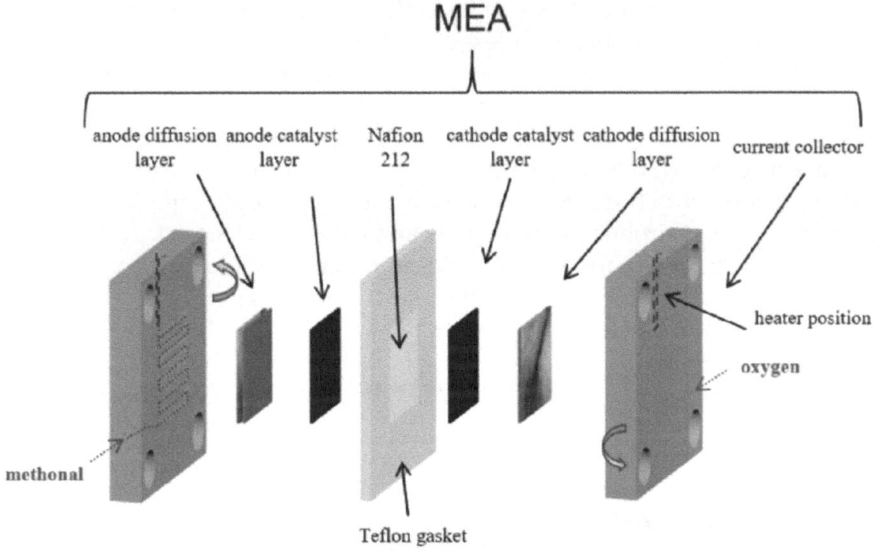

FIGURE 3.3
Schematic diagram of direct methanol fuel cell fixture [33].

According to research, the total performance of DMFCs is significantly affected by the flow field design, which includes the flow field profile, depth, breadth, and length. Good CO_2 and water removal systems, minimal methanol crossover, high mass transfer, and low concentration fluctuation in each cell contribute to outstanding performance. An increase in the partial pressure of oxygen inside the channel may improve a cell's performance and operational stability, and shallower channels can improve the mass movement of air and water. The more significant pressure drops in shallower channels, particularly with a prolonged cell, contribute to higher pumping needs [35].

6. Gas diffusion layer (GDL): There is a simultaneous release of water into the cathode and diffusion of reactant methanol from the methanol reservoir to the catalyst layer at the GDL. The most popular GDL material is carbon, which meets the requirements of being both porous and conductive [36].
7. Gasket: A gasket keeps the reactant from leaking out and separates the current collector from the end plate. The gasket's thickness affects the performance, which increases the internal resistance. A more extended DMFC system with a thicker gasket has a higher total internal resistance. Silicon rubber is the material most often used for gaskets.
8. Current collector and end plate: The current that is produced in the MEA is collected by the current collector in a DMFC system. A homogeneous transport area, excellent mechanical strength, and high electric conductivity are all necessary characteristics of the current collector. Some current collectors have a low effective thermal conductivity, which helps to keep the temperature stable by reducing heat loss. Copper, stainless steel, gold-plated stainless steel, platinum-coated niobium, and copper with a gold plating have all been used by researchers as current collectors [37].

The additional DMFC components are housed between the end plates, which, in conjunction with bolts, help to sustain and keep the system compact. The material should be robust mechanically and not conduct electricity. Stainless steel and other metals are common choices for end plates. However, a gasket to isolate the collector from the endplate is frequently necessary.

Though they are generally related to PEM fuel cells, DMFCs are unique in that they run on liquid methanol instead of gasoline. In this part, we'll look at the advantages and disadvantages of DMFCs compared to other common fuel cell types.

3.2.1 Hydrogen Fuel Cells (PEMFCs)

Compared to other fuel cell types, PEMFCs stand out for several reasons, including their low-temperature operation (often between 80 and 95°C), high power density, low maintenance cost, quick start-up, and high theoretical efficiency (for hydrogen fuel) of 83%. Proton electrolyte fuel cells (PEMFCs) use more straightforward MEA configurations than other fuel cell types. The required voltage and current may be produced by linking the fuel cell stack in either series or parallel. The MEA consists of the cathode and anode, a proton-conducting membrane (Nafion or perfluoro sulfonated membrane), and catalyst and porous gas diffusion layers. However, the cost of H_2 production, transport, and storage in high-pressure tanks is increasing despite hydrogen having a high energy density per weight. DMFCs are advantageous because they use liquid methanol instead of hydrogen gas, which is more convenient for storage and handling. Fuel logistics are made more accessible because methanol may be kept at room temperature and pressure [38].

3.2.2 Solid Oxide Fuel Cells

Solid oxide fuel cells (SOFCs) employ non-noble catalysts and operate in the 600–1000°C range with solid nonporous metal oxide as the electrolytes. Fuels like MCFCs self-reform at high temperatures. The most typical electrolyte for SOFC systems is yttria-stabilized zirconia (YSZ), which enhances fuel oxidation by combining ionic conductions with noble metals; SOFCs have an efficiency range of 50 to 60% when using this electrolyte. SOFCs are suitable for a wide range of micro to several MW power applications because of their durability and the excellent quality of the heat they generate due to the combined cycle. Unfortunately, fuel cell materials have poor durability in very hot conditions. Several hydrocarbons and natural gas products are among the fuel options for SOFCs. Because of their high operating temperatures (600–1000°C), SOFCs can reform hydrocarbon fuels internally and reach very high efficiency levels. Since DMFCs only work at temperatures between 60 to 120°C, they are ideal for portable and small-scale applications. Because the working temperature is lower, the cell components have less thermal stress and fewer starting periods [39].

3.2.3 Alkaline Fuel Cells

Alkaline fuel cells (AFCs) utilize electrolytes, which may be concentrated solutions of materials with a high pH, such as potassium hydroxide (KOH). Their operating temperature range is 100 to 250°C, and they are the most economical fuel cells. Although nickel is the most common catalyst used in electrochemical reactions, studies have shown that porous electrodes may enhance the kinetics of many other metals and noble metal oxides. AFC eliminates the bipolar plates, and the remaining water, power, and heat might be used in other noteworthy applications, such as crewed spacecraft (Shuttler and Apollo). Its key advantages are the strong oxide reduction reaction at the anode in an alkaline medium, low operating costs, and high electrical efficiency (around 60%) of AFCs. The catalyst might be poisoned by CO_2 in AFC, which could significantly affect performance. KOH and potassium carbonate (K_2CO_3) react because of CO_2 in AFCs. An alkaline electrolyte is used to power AFCs, which run on hydrogen as their fuel (usually KOH). Because of their dependability and efficiency, AFCs are primarily used in space applications. Despite their efficiency, AFCs are vulnerable to carbon dioxide, which may cause the electrolyte to deteriorate via the formation of carbonates. DMFCs are free of this problem and can run on a carbon-neutral fuel source renewable methanol [40].

Compared to other fuel cell technologies, DMFCs have some benefits, especially in the following areas: Because it is a liquid at room temperature, methanol is convenient to store, transport, and replenish because of how simple it is to handle. Contrast this with hydrogen, which must be stored at freezing or under extreme pressure. Since methanol has a greater energy density per unit volume than hydrogen, DMFC systems are well-suited to small, lightweight, portable applications [41]. Due to its scalability, DMFCs may be reduced in size without sacrificing efficiency, making them perfect for backup power systems and portable devices. Methanol can be created from renewable resources, including waste materials, biomass, and carbon dioxide capture methods, which means it might be a fuel that doesn't contribute to climate change. To fully grasp the energy landscape's potential for DMFCs, it is essential to understand their basics. A clean and efficient alternative to existing power sources based on fossil fuels, DMFCs transform the chemical energy of methanol directly into electrical energy via electrochemical processes [42]. The many applications that DMFCs are well-suited for are highlighted by their comparative benefits, significant components, and operating principles.

Fuel Cell	Fuel	Operating Temperature	Electrolyte	Power Density	Efficiency	Cost
AFC	hydrogen	70–90	alkaline electrolyte	~ 0.5	~ 70	high
DMFC	methanol	60–120	polymer electrolyte membrane	~ 0.5	~ 40	low
PEMFC	hydrogen	60–100	polymer electrolyte membrane	~ 4	~ 60	high
SOFC	hydrogen natural gas, biogas	500–1000	ceramic electrolyte	~ 1.5	~ 60	high

3.3 Technological Advancements in DMFCs

Performance, durability, and economic viability have been achieved via substantial technical improvements in DMFCs. Here, we look at the recent advances in DMFC technology, including new catalysts, membrane materials, and cell designs. Because they enable the electrochemical processes at the anode and cathode, catalysts are vital to the efficiency and performance of DMFCs. As a result of their substantial catalytic activity, conventional catalysts, like platinum (Pt) and platinum-ruthenium (PtRu) alloys, have long been used. These materials aren't cheap and will eventually deteriorate. Creating longer-lasting and less expensive catalysts has been the primary focus of recent innovations [36,43]. Methanol is electro-oxidized at the anode of DMFCs. Polarization losses mainly cause the slow reaction kinetics at the anode. Platinum catalysts are still the best option for DMFC anode catalysis, even though they are costly. Consequently, platinum catalysts have been the primary focus of research into the oxidation process of methanol [44]. The following procedures make up the MOR mechanism: Various intermediates, such as HCHO, HCOOH, CO, and so on, are formed during the adsorption and de-protonation of methanol molecules. The significant binding strength of these intermediates allows them to block the active sites of catalysts, which negatively impacts MOR, resulting in poor cell performance and a shorter catalytic life lifetime. Consequently, converting these intermediates to CO_2 requires an active catalyst. As a result of intermediate oxidation, carbon dioxide is produced. Finally, CO_2 is released, and electrons are transferred across the external circuit [45]. The anode of PEMFCs produces the HER. Hydrogen undergoes oxidation here, releasing electrons that create protons. After that, the electrons go via the external circuit to make electricity. A typical catalyst for HER at the anode is platinum on carbon support (Pt/C); nevertheless, Pt/durability is still a concern. An irreversible degradation of platinum metal begins during transient fuel cell operations (start-ups, shutdowns). It impacts the assembly of electrons in the membrane, leading to a drop in the fuel cell's performance output. In light of this, there have been many efforts to find solutions to the durability problem by researching and finding new materials. For example, new research found that a 1:12 alloy of iridium and ruthenium on a carbon substrate might achieve the same catalytic performance as traditional Pt/C catalysts, with 120 times higher durability. Research findings have shown that magnetron deposition of carbon in a nitrogen environment may modify carbon supports, making them more robust and corrosion-resistant. This modification is used to develop Pt catalysts. In turn, this makes electrodes that operate similarly last longer. This has a negligible impact on the overall cost of the PEMFC stack since Pt/C is often employed as an anode catalyst with a much smaller loading than when used as a cathode [46].

A nanoporous thin film (NPTF) was used to deposit compressively strained PtRu alloy nanoparticles. A differential normal pulse voltammetry approach was used for this purpose. According to geometrical phase analysis, which shows high-level compressive stresses of around −5%, the resultant compressive structure is supported (Figure 3.4). The extensive specific surface area and enhanced specific activity of the compressively strained

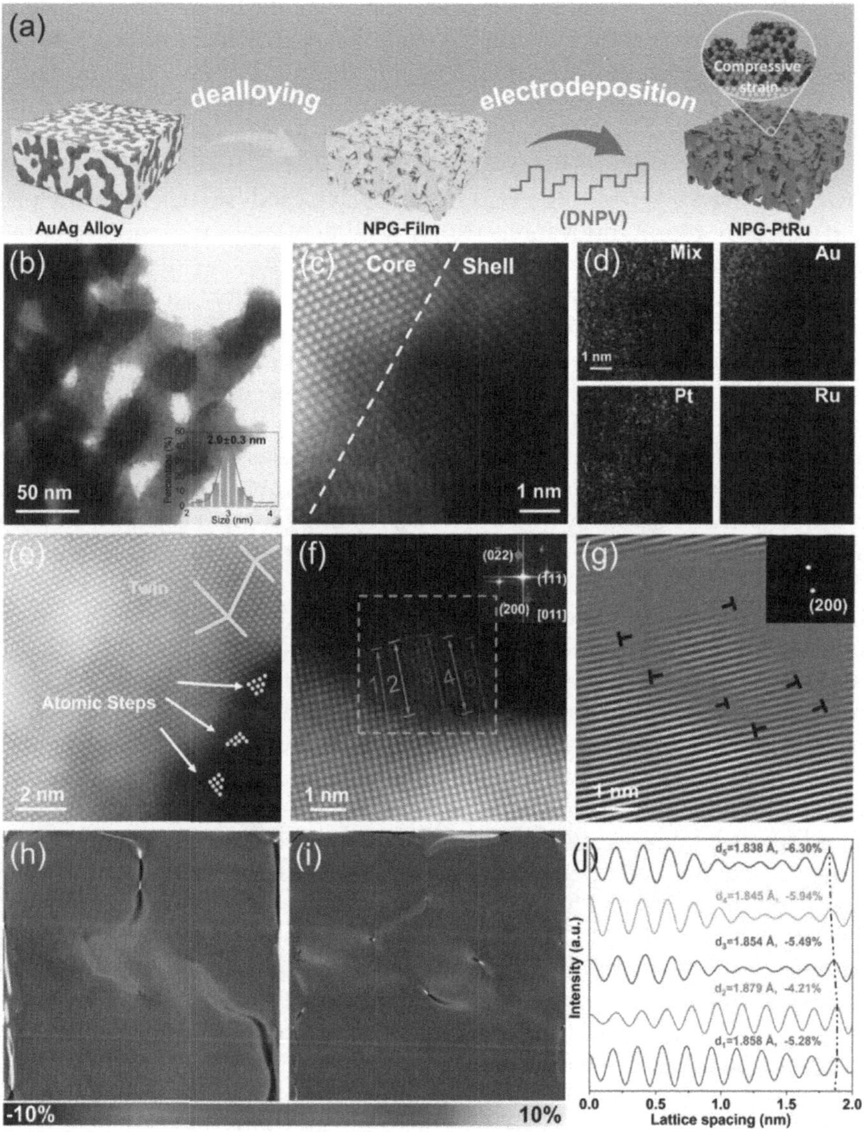

FIGURE 3.4
Structural characterization of NPG-PtRu. (a) Schematic picture of the synthetic procedure for NPG-PtRu samples. Typical TEM (b) and HAADF-STEM (c) images of the NPG-Pt_2Ru1 electrocatalyst, where the white dashed line in the image (c) denotes a possible boundary between the core and shell. (d) EDS elemental mapping images of Au, Pt, Ru, and their overlay according to image (c). (e and f) Atomic resolution HAADF images of selected regions of NPG-Pt_2Ru1 (the inset shows the corresponding FFT pattern of (f)). (g) Corresponding iFFT patterns of the (200) plane from (f). (h and i) Strain distributions of exx and eyy corresponding to the HRTEM image in (f). The compressive strain is represented by green to dark blue, while the compressive strain is depicted by red to bright yellow; the colour bar indicates −10 to 10% strain. (j) Corresponding intensity profiles of the lattice spacing of Pt_2Ru1 nanoparticles taken from different regions in (f) [47].

PtRu alloy nanoparticles make them very active in methanol oxidation. At 1.64 A $mgPt^{-1}$, NPTF-Pt_2Ru_1 has the most excellent Pt mass activity, over four times more than commercial PtRu/C. Due to structural benefits, such as attenuated methanol crossover, this film anode further demonstrates mass activity augmentation when employed in membrane electrode assembly in DMFCs. At 80°C, the highest power density approaches 117 mW cm^{-2} with a loading of 0.21 mgPt cm^{-2}, and the power efficiency of Pt is almost 12 times more than that of commercial PtRu/C (2 mgPt cm^{-2}, 92 mW cm^{-2}). As a result, the study's technique shows a new and effective way to make DMFC anode catalysts that work well [47].

A greater specific surface area, which is crucial for methanol oxidation, was achieved by anodizing surfaces of highly ordered porous titanium dioxide nanotubes (TiO_2-NTs). Here, ideal circumstances were identified, including different anodization voltages and durations. The maximum TiO_2 surface area was achieved at an anodization voltage of 60 V and a duration of 2 hours. The behavior of the nickel nanoparticles during methanol electro-oxidation was studied using cyclic voltammetry (CV), electrochemical impedance spectroscopy (EIS), and chronoamperometry (CA) after a TiO_2-NT surface with a high specific surface area had been prepared. The SEM scans show that the average diameter of the tube is 93.6 nm and the average length is 36.32 μm. TiO_2 and Ni (111) anatase with a tetragonal structure was shown by XRD analysis (200). Although the TiO_2-NT surface does not show a peak for methanol oxidation, the behavior of methanol oxidation is affected by the Ni level on the surface. As the concentration of Ni nanoparticles in the deposits rises, the oxidation response also increases. The Ni-modified TiO_2-NT with the most excellent Ni content has a deficient activation energy (Ea) of 11.0 kJ/mol and a high surface coverage (Γ) of 3.87×10^{-9} mol cm^{-2}. When Ni is deposited over TiO_2-NT, the charge transfer resistance decreases or enhances, leading to extended stability and durability. This suggests that the high surface area of TiO_2-NTs may be a critical factor in the efficiency of methanol oxidation. Adding Ni particles to the surface of TiO_2-NT is a great way to improve the electrocatalytic performance. In addition, the bifunctional mechanism on the surface of the electrode may make it easier for methanol to be adsorbed due to the solid electrical contact between Ni and TiO_2 [48].

Noble metal alloys are the catalysts of choice when oxidizing methanol. Nevertheless, the production of noble metals is a complex and costly process. For this reason, there has been a concerted attempt to lower catalyst alloys' noble metal concentration. Here, PtCox nanoparticles enclosed in carbon nanotubes are synthesized using a straightforward g-C_3N_4-assisted thermal approach ($PtCo_x$ CNTs). Thanks to their distinctive carbon structure, well-calibrated active site ratio, and appropriate electronic states, $PtCo_3$ CNTs have the most significant catalytic activity in MOR among all the $PtCo_x$ CNTs catalysts studied (x = 1, 1.5, 3, and 6). The $PtCo_3$ CNTs outperform commercial Pt/C by a factor of 4.4 in specific activity and 7.7 in mass activity. The stability of $PtCo_x$ CNTs is improved by adding more Co, and the electrochemical test shows that the catalytic activity of $PtCo_3$ CNTs is almost constant even after 1,500 cycles. In a prototype methanol fuel cell, the $PtCo_3$ CNTs show any catalyst's greatest power and current densities. This research details an easy way to make MOR-active, Pt-based alloy nanoparticles encased in carbon nanotubes with a low Pt ratio [49].

3.3.1 Cathode Materials

The electrocatalytic activity, Pt loading, and total cost may be improved by alloying platinum with transition metals. The surface of a Pt-M (where M is a metallic species) catalyst may be either (a) randomly mixed with homogeneous or heterogeneous M and Pt distribution, (b) M surface with pure Pt atoms, or (c) Pt-skeleton surface, depending on the arrangement of transition metals in this surface. The Pt alloys' superior performance in terms of strain and the ligand effect has been confirmed. Compression is applied to the surface layers when there is a size disparity between the surface and the atoms close to it, leading to the strain effect. The d-band centre downshifts during compressive strain, leading to poor binding of reaction intermediates and enhanced efficiency; in contrast, significant adsorption of reaction intermediates occurs under tensile strain, which shifts the d-band centre near the Fermi level. The ligand effect changes Pt's interaction with reaction intermediates and its electrical characteristics. While both the strain and ligand effects are interdependent, the former takes precedence. Iron, a plentiful and inexpensive metal, might be a perfect match for platinum alloy catalysts, as cost is the primary obstacle to commercializing DMFCs [50].

In direct methanol fuel cells, methanol crossover is a significant issue. At the cathode, the MOR poisons the cathode catalyst and degrades the DMFC performance as it passes through the membrane from the anode to the cathode (Figure 3.5). This highlights the critical need for a cathode catalyst to promote the ORR, a common cathode reaction in DMFCs while suppressing MOR. To that end, our research created a Pt cathode catalyst encased in carbon. Carbon nanofibers coated with a Pt-aniline complex were heat treated to produce the catalyst. As the graphitization degree of the catalyst's carbon shell grew, it became more apparent that it effectively blocked methanol from reaching the Pt core. At the same time, the Pt core could participate in ORRs since the carbon shell lets O_2 through, independent of the degree of graphitization. Compared to conventional Pt/C, the synthesized catalyst performed better and was more stable in single-cell experiments conducted under various circumstances [51].

DMFCs based on Pt/C always have a significant crossover when exposed to solutions with high methanol concentrations. A new hybrid cathode catalytic layer (CCL) made of a physical mix of Pt/C and Fe-N-C is described in this work (Figure 3.6). At 10 M methanol, the DMFC based on the hybrid CCL with optimal loadings (0.25 mg cm^{-2} Fe-N-C and 2 mgPt cm^{-2} Pt/C) produces a high-power density of 28.87 mW cm^{-2}, whereas the single Pt/C based cell's

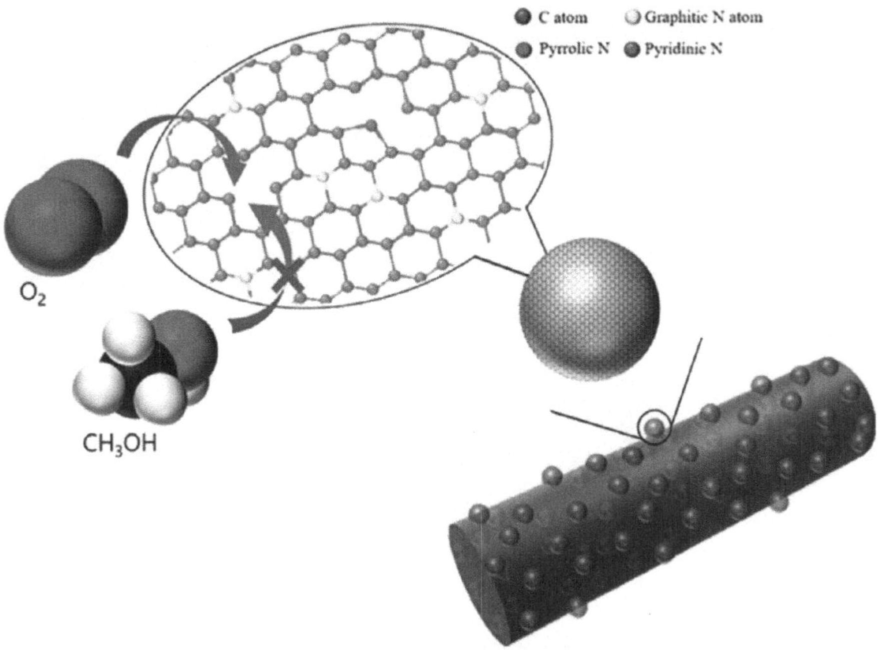

FIGURE 3.5
Carbon-encapsulated Pt cathode catalyst with reduced methanol crossover [51].

power density is only 12.04 mW cm^{-2}. Fe-N-C abundantly creates reactive oxygen species (ROS) via an imperfect four-electron transfer pathway in oxygen reduction reactions, which is responsible for the significant improvement. Cu-N-C demonstrates the most impressive performance in improving the methanol tolerance of Pt/C out of various M-N-C catalysts inspired by this technique. At 10 M methanol, the peak power density of DMFC cells using Cu-N-C hybrid CCL surpasses 31 mW cm^{-2}, making them superior to Fe-N-C hybrid CCL cells. This study offers a novel method for fabricating high-energy-density DMFCs with real-world implications for portable power device applications [52].

An important aspect of DMFCs is their efficiency in the ORR. To study the function of CeO_2 in ORRs, this study synthesized composites of MnO_2 supported on CeO_2-QDs (CeO_2-QDs@MnO_2) and CeO_2-QDs supported on MnO_2 (MnO_2@CeO_2-QDs), using varying loading methods (Figure 3.7). CeO_2 quantum dots are known for their excellent oxygen adsorption and ion conductivity (CeO_2-QDs). The components of a passive DMFC were a cathodic catalyst of CeO_2-QDs@MnO_2 or MnO_2@CeO_2-QDs and an anodic catalyst of PtRu/C. The greatest power densities for DMFCs based on CeO_2-QDs@MnO_2 were 9.4 and 7.8 mW·cm^{-2}, respectively, whereas for DMFCs that breathed air, the values were 8.5 and 7.0 mW·cm^{-2}. In particular, for air-breathing DMFCs,

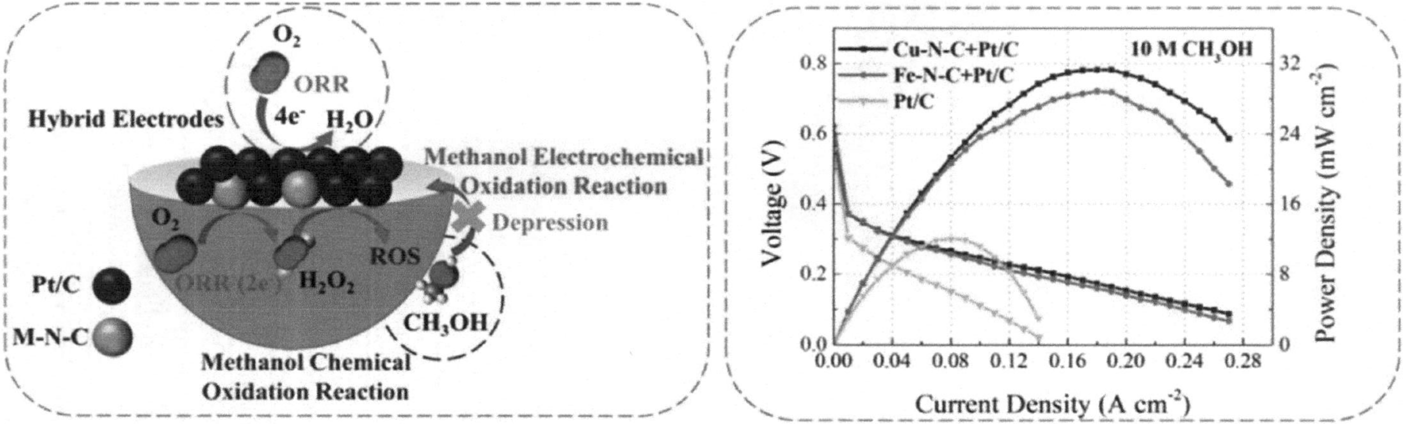

FIGURE 3.6
(a) Schematic diagram of the depression of MEOR by MCOR of ROS and the enhanced ORR at the hybrid electrode composed of Pt/C and M-N-C. *(b)* Polarization curves of DMFCs with different loadings of Pt/C, Fe-N-C+Pt/C, and Cu-N-C+Pt/C at CCLs fed with 10 M (b) methanol solution [52].

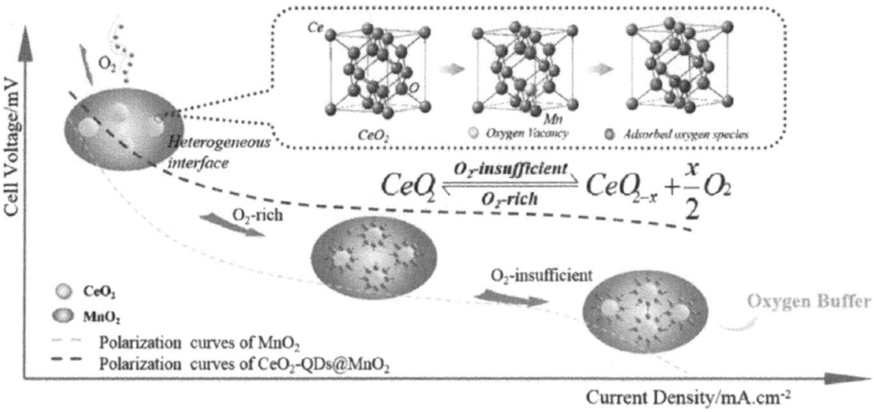

FIGURE 3.7
Catalytic mechanism schematic illustration for CeO_2-QDs@MnO_2 catalyst [53].

CeO_2-QDs@MnO_2 demonstrated better ORR activity and stability at high discharge currents. The ORR efficiency at the cathode might be dramatically improved by using the CeO_2-QDs@MnO_2 catalyst, which has a mechanism for temporary storage and prompt oxygen replenishment. A novel approach of loading CeO_2 onto CeO_2/MnO_2 catalysts for ORRs was proposed in this study [53].

One of the most critical parts of PEMFCs, apart from catalysts at the electrodes, is the electrolyte, which is situated between the anode and the cathode. PEMs utilized in DMFCs typically include Nafion, fluorinated composites, and nonfluorinated membranes. Research indicates that DMFC membranes should have favorable properties, such as minimal fuel crossover, excellent thermal and economic stability, and strong proton conductivity [54]. Ion movement inside DMFCs, from anode to cathode, is mainly the responsibility of the PEM, a critical component of these cells. The function of PEM extends beyond ion transportation to include electrode separation, cell stability, and power output delivery. Due to its high ionic conductivity and remarkable stability, Nafion has been studied as a potential electrolyte for perfluorosulfonic membranes used in DMFCs [55].

Nevertheless, the need to explore other membrane materials for DMFC has resulted in a rise in cost since catalyst poisoning at high temperatures and higher methanol crossing have degraded Nafion. Several studies have shown composite membranes that perform as well as Nafion in this regard. Nafion is the best membrane for this purpose, a tetrafluoroethylene-sulfonate copolymer with outstanding proton conductivity (0.1 S cm^{-1}) and excellent mechanical and thermal durability [56]. However, we are always looking for better alternatives due to drawbacks, such as high fuel usage, high processing and manufacturing costs, hostile eco-nature (hazardous production materials), and reduced thermal stability at higher temperatures. Several alternatives have been investigated for potential use in DMFCs, including chitosan (CS), sulfonated polyether (ether ketone) (SPEEK), non-sulfonated and sulfonated aromatic polymers (SAPs), polybenzimidazole (PBI), sulfonated polyethersulfones, and many more [30,45,57].

The authors present a composite proton exchange membrane for DMFCs made of sulfonated polyether ether ketone (SP/CT-MOF-3) and a MOF based on copper-benzene tricarboxylic acid (Cu-TMA) (Figure 3.8). The composite membrane of MOFs is very resistant to methanol crossover in DMFCs. The MOF effectively blocks SPEEK pore channels without influencing its proton conductivity, reducing methanol permeability. Permeability values for methanol in SP/CT-MOF-3 are 4.26×10^{-7} cm^2 s^{-1}, whereas in virgin SPEEK, they are 7.95×10^{-7} cm^2 s^{-1}. Compared to pure SPEEK, SP/CT-MOF-3 has a superior proton conductivity of 45 mS cm^{-1} and an activation energy of 7.27 kJ mol^{-1}. SP/CT-MOF-3 outperforms virgin SPEEK in peak power density (116 mW cm^{-2}) and current density (570 mA cm^{-2}). Due to its superior ion exchange capacity, water sorption, and proton conductivity compared to pristine SPEEK, the SP/CT-MOF-3 composite membrane electrolyte outperforms its DMFC counterpart [58].

DMFCs rely on fabricating new PEMs, and a simple in-situ growing method is used to attach molybdenum disulfide (MoS_2) nanosheets onto carbon nanotubes (CNTs). Subsequently, a SPEEK matrix was used to build a variety of composite PEMs by embedding MoS_2-coated CNTs (MoS_2@CNTs) (Figure 3.9). Coating the composite membranes with MoS_2 nanosheets increased their mechanical characteristics, made them more dispersible, and made them more compatible with the polymer matrix. Composite membrane ionic conductivity might be improved by building channel-like ion transport routes (water channels) from CNTs with a hydrophilic MoS_2 coating, thanks to their unique one-dimensional structure. Noteworthy is the fact that a 1.7-fold proton conductivity compared to a pristine SPEEK membrane and drastically reduced methanol permeability were produced by the 80°C inclusion of 1 wt percent MoS_2@CNTs. Consequently, the SPEEK/MoS_2-1 composite membrane outperformed the SPEEK and Nafion-115 membranes in fuel cells, boasting an impressive peak power density of 98.5 mW cm^{-2} at 70°C. Lastly, the durability test further confirmed the promise of the SPEEK/MoS_2@CNTs composite membrane as a Nafion membrane replacement [59].

3.3.2 System Design

While planning the framework's materials and components is essential, the state of technology is such that executing those materials and components constrains the selected framework level. The following is a summary of the specific challenges that have arisen along the process of bringing these devices to market: sensitivity to disturbances and directions, fuel production, electrochemical activity, and heat and mass adjustment. Improvements and choices in catalyst layer, bipolar plate, electrode, and PEM material selection

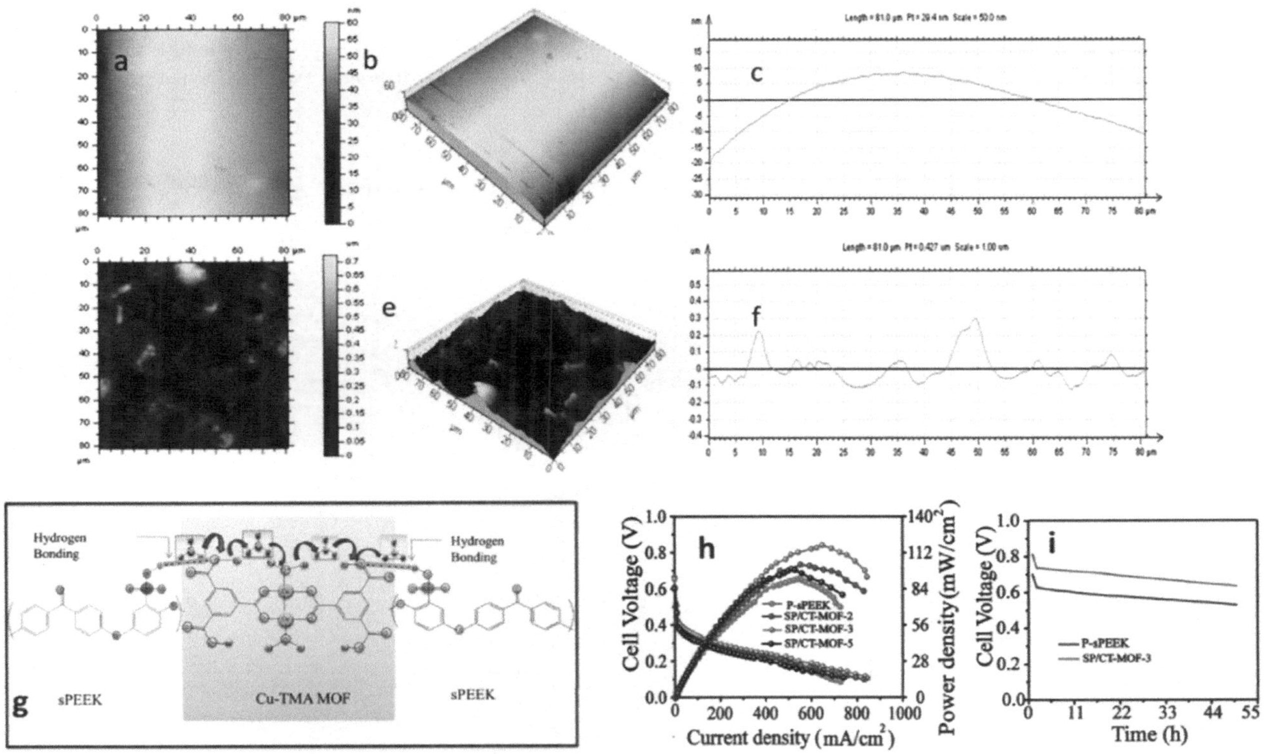

FIGURE 3.8
AFM analysis for pristine SPEEK (a and b) and the SP/CT-MOF-3 (d and e) composite membrane. Roughness profiles for both membranes are seen in (c and f) (g).

Source: Illustration of proton transport through Cu-TMA MOF in the composite membrane. (h) DMFC cell polarization and (i) stability data representing OCV as a function of time for pristine SPEEK and SP/CT-MOF composite membranes [58].

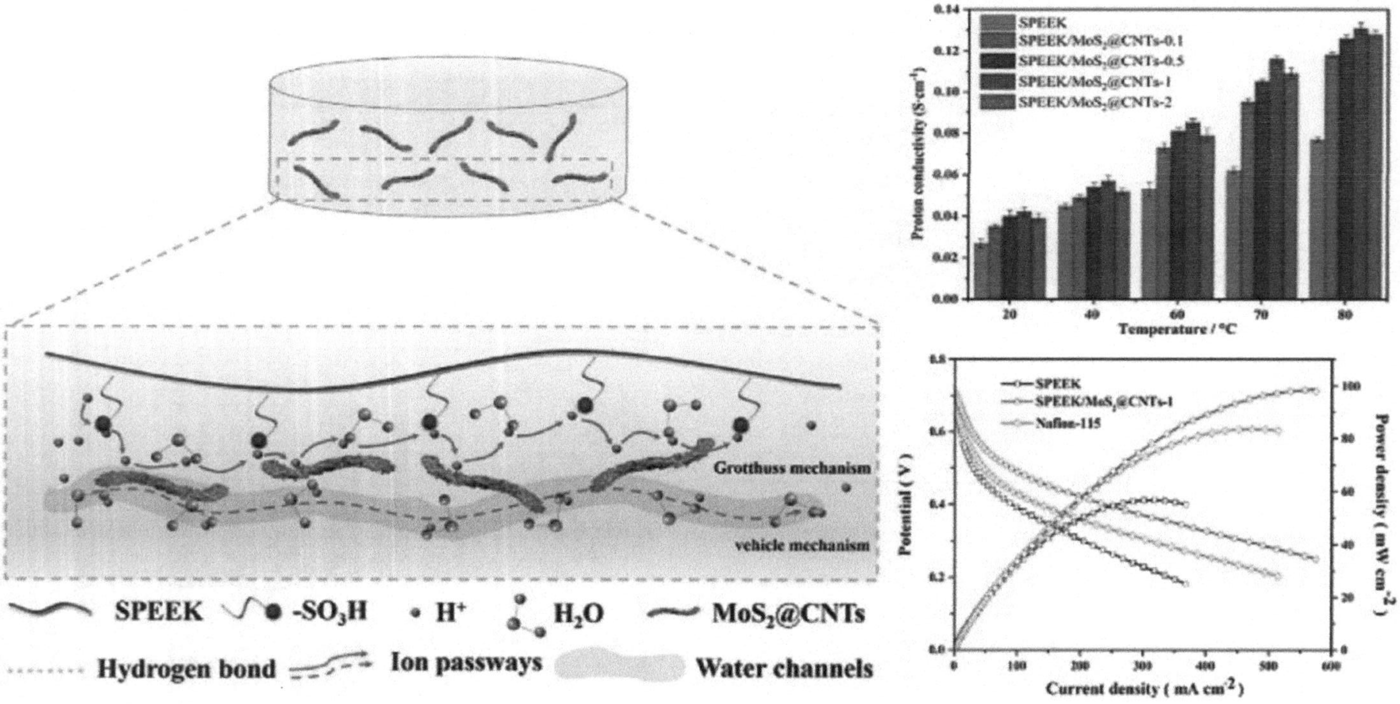

FIGURE 3.9
(a) Schematic diagram of hypothesized proton transfer mechanism in SPEEK/MoS_2@CNTs composite membranes. (b) proton conductivity of membranes at different temperatures. (c) Power density and polarization curves of DMFCs using pristine SPEEK, SPEEK/MoS_2@CNTs-1, and Nafion 115 membrane with 2 M methanol at 70°C [59].

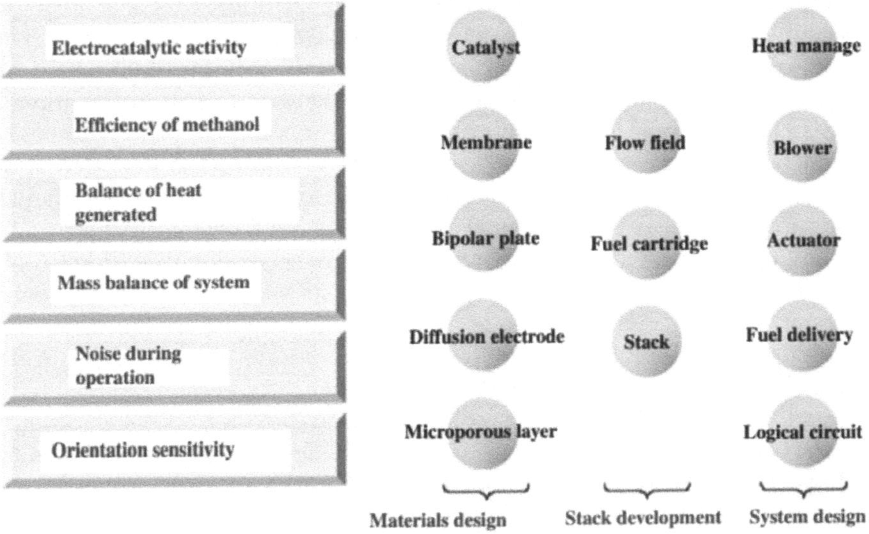

FIGURE 3.10
Technological roadblocks involved in commercializing DMFC [60].

may manage most of these difficulties. Obstacles to commercializing DMFCs are shown in Figure 3.10. First, the cost of materials; second, the life cycle of the product; third, the production method; fourth, the need for standardization; fifth, customer habits; and sixth, developments in mobile devices [60].

3.4 Environmental and Economic Benefits

The prices, emissions, recycling possibilities, toxicity of methanol and its manufacturing, and the shift to DMFC technologies will have social and environmental consequences [61]. Furthermore, substituting methanol for other fuels will have substantial social and environmental repercussions. The quantity and kind of fuel that is replaced and the amount of methanol needed by the implemented technology will determine the extent to which these effects are felt. Adopting DMFC technologies that use comparatively less methanol will have far-reaching environmental consequences, unlike technologies that require vast volumes of methanol, such as cars (like consumer electronics) [62].

3.4.1 Environmental Toxicity of Methanol

Methanol has already made its way into the environment due to many different processes, both natural and manufactured. Under typical conditions, methanol has little to no environmental impact due to its low toxicity. Since methanol breaks down quickly in aerobic and anaerobic settings via photo-oxidation and biodegradation, it is not known to build up in soil or water sources. As methanol usage rises, the likelihood of an unintentional release of methanol into the environment during manufacturing, transportation, or storage grows, as is the case with any similar fuel source. Methanol has many significant benefits over conventional fossil fuels, even though fuel leaks are never ideal. In the event of a spill, methanol's low toxicity and quick decomposition provide a distinct advantage over gasoline or diesel fuel, including several harmful and carcinogenic substances that break down extremely slowly. Also, if there is a big spill into the water, the methanol would quickly be diluted and dissipated to harmless levels since it is entirely miscible with water. Microbes in the environment would then rapidly break down the leftover diluted methanol into carbon dioxide and water. Several recent oil spills have shown that oil is insoluble in water, which leads to extensive and long-lasting pollution and, in many cases, catastrophic effects on the surrounding ecosystem. These benefits contribute to methanol's reputation as a cleaner, safer fuel option [63].

Worldwide, people are worried about emissions from vehicles and their engines since these pollutants often include toxic levels of volatile organic compounds (VOCs), particulate matter (PM), sulfur dioxide (SO_2), carbon monoxide (CO), and other nitrogen oxides (NO_x) [64]. The dangerous nature of these pollutants has prompted several regional, national, and worldwide regulatory agencies to impose ever stricter standards on transportation emissions. A number of alternatives are now being investigated, including DMFC vehicles, as a result of these regulations' push for more sustainable transportation options [65].

DMFC cars are anticipated to be far better than the average gasoline or diesel internal combustion engine (ICE) in terms of emissions since they will be zero-emission vehicles (ZEVs). Hybrids, electric cars, hydrogen-powered fuel cell vehicles (H_2-FC), and other alternatives to conventional ICE vehicles all contribute to lower emissions [66,67]. Regarding direct (tailpipe) emissions, both H_2-FC and EVs are designated ZEVs, which puts them on a level with DMFC cars. Using a mix of battery power and internal combustion, hybrid vehicles can reduce exhaust pollution by about 80% compared to similar ICE vehicles.

The progressive adjustments to emission requirements have allowed for more straightforward technological solutions to be implemented while reducing emissions to a certain extent, notwithstanding the apparent benefits of DMFCs and other alternative technologies. Combining methanol with gasoline for *FFVs* or utilizing pure methanol fuel (M100) would be one straightforward way to use methanol in an ICE, reducing emissions. Aside from producing less VOCs and nitrogen oxides than gasoline, burning methanol does not produce particulate matter. Substituting M100 for gasoline in ICEs may lower hydrocarbon emissions by as much as 80%, according to the EPA; however, actual emissions would vary depending on engine type. Despite methanol's inherent cleaner combustion, it's doubtful that it would significantly drop emissions compared to gasoline. Emission-control technologies continuously improve and significantly reduce emissions, making conventional fuels competitive with methanol and other alternatives. Any shift in fuel or technology will rely heavily on infrastructure, even though transportation-related emission reductions are a critical component. Gasoline and diesel fuel are the backbone of the transportation sector's extensive infrastructure. Methanol and DMFC cars have an edge regarding how easily they integrate into the current fuel infrastructure system, which is a key factor in adopting any alternative fuel or technology. Since methanol is a liquid fuel, it may be utilized in the same way that gasoline and diesel fuel are used in vehicles now, with some tweaks made to the fuelling systems and onboard storage. Existing double-walled underground storage tanks for gasoline or diesel fuel would also need to be converted. New piping and dispenser pumps that can handle methanol must be installed at fuelling stations [68].

Fuelling infrastructure upgrades tailored to DMFC vehicles are very improbable; nevertheless, a more plausible possibility is the launch of a methanol-based FFV, which might shift the tide and start the transition to methanol-based fuel. Accordingly, switching to DMFC cars may be more easily pulled off in the long run if methanol-ICE FFVs are adopted shortly.

Two components determine the final price of fuel cells: production expenses and consumer spending. Costs incurred by the end user are only 14% of the whole cost. The remaining cost of fuel cells comprises manufacturing costs, including design, materials, assembly, and capital equipment costs. The electrodes are the primary contributors to the system cost, making up about half of the stack cost. For instance, DMFCs rely heavily on the rare and costly platinum for their electrodes.

Consequently, the price of DMFC has gone up. Conversely, research has shown that reducing platinum loading may affect fuel cell durability. However, it is paramount to find an excellent medium between fuel cell prices and their longevity.

Since current technologies (batteries and oil/gasoline) are more economically viable, DMFCs have not yet made a big splash in the market. Several policy measures may encourage transitioning to technologies based on DMFCs, which are alternatives to traditional fossil fuels. While some governments pass laws to promote the development of specific technologies, others are more interested in opening the door to various innovations within a single field (i.e., clean energy). A few policy options that could affect DMFC technologies are cutting fossil fuel subsidies (currently around $550 billion per year), increasing R&D spending (generally for clean energy technologies but also specifically for DMFC technologies), tightening regulations on emissions from industries and vehicles, creating a tax credit for CO_2 or other emission reductions, or even a carbon, emissions, or energy tax.

While most industrialized nations invest in clean energy R&D, the quantity and distribution of this investment vary greatly from one nation to the next and even from one year to the next due to fluctuations in spending and technological advancement. Similarly, most industrialized countries have enacted emission rules, with many of these laws requiring drastic cuts to emissions over ten to 50 years. Even though emission regulations do not directly affect DMFC technologies, these methanol-based solutions can still benefit from them because DMFC vehicles offer emission reductions compared to traditional gasoline-ICE vehicles and methanol production itself can reduce emissions.

The environmental and economic benefits of DMFCs make them a promising technology for the future of sustainable energy. Their ability to reduce greenhouse gas emissions, lower pollutants, and utilize renewable methanol positions them as an attractive alternative to traditional fossil fuel–based power sources. Economic viability is enhanced by ongoing catalyst and membrane technology advancements, reducing costs and improving performance. The potential for DMFCs to contribute to a carbon-neutral energy system is significant, mainly when renewable methanol production pathways are fully realized. As we continue to explore and develop DMFC technology, its role in the energy transition and sustainability landscape will undoubtedly grow.

3.5 Challenges and Limitations

The benefits of DMFC over other fuel feed cells and its promising future applications are highlighted by its current advancement. The DMFC system has great potential. However, it is currently limited in its ability to be commercialized and developed further. If all existing obstacles are quickly and effectively overcome, DMFC may be a viable option for producing renewable energy sources. Battery technology and fossil fuels are formidable competitors to DMFCs in the industrial sector. In this part, we will primarily focus on the present issues each form of DMFC faces. Also included as a road map for researchers to follow is a suggestion for overcoming these obstacles.

Dealing with the expense of catalysts is the first hurdle for DMFC. Additionally, DMFC performance is highly dependent on high catalyst loading. Therefore, compared to other systems, such as lithium-ion batteries, DMFC faces significant obstacles due to the simultaneous occurrence of the issues mentioned. A more substantial catalyst loading percentage is required in DMFC systems than in hydrogen PEMFC systems, and all DMFC types rely on catalysts based on noble metals. Both the anode and cathode sides of the DMFC system employ catalysts made of PtRu and Pt, respectively, in commercial applications. A catalyst dosage of 8 mg cm^{-2} is necessary for both electrodes to achieve a maximum power density of 23 mW cm^{-2}. Many researchers have looked for ways to lessen the catalyst loading.

In addition, DMFC has difficulties due to fuel crossover, which happens when fuel moves from the anode side to the cathode side via the membrane separator. Because there is a concentration gradient in liquid fuel on each side of the electrode, transferring fuel from high- to low-concentration regions, the fuel crossover problem is more severe in liquid fuel feed systems than in gas fuel feed ones (anode to cathode). To generate electricity efficiently, fuel molecules must oxidize to the anode side, but some fuel is lost as they cross the membrane layer. Furthermore, oxidation generates CO, poisoning the Pt catalyst. As a result of electro-osmosis strain, fuel diffusion, and absorption, fuel cross-transfer might occur. Anode catalysts with efficient fuel oxidation, concentration management of fuel supply on the anode, and improved proton exchange membrane performance may reduce this cross-linking issue. Matrix composition, structure, and thickness of the fuel transfer membrane, as well as cell operational limitations, such as fuel concentration and temperature, have a role in the cross-linking of fuel transfer. Oxidation of methanol molecules occurring on the positive electrode due to diffusion across the membrane is known as methanol crossover. An extremely critical issue that drastically lowers cell performance is methanol crossover, which occurs when methanol moves from the anode to the cathode. Except for the anode, the remaining 70% of the chemical energy in methanol is lost to heat due to methanol crossover and the irreversibility of electrode reactions, leaving just 30% usable as power. To decrease methanol crossover, it is necessary to dilute methanol despite its high energy density (about 1.8 kWh kg^{-1}). The reduction in concentration necessitates an increase in the size of the cell stack in direct proportion. Given the cell's intended use in portable electronic devices, it is crucial to accurately predict the power consumption in response to its rising size. Secondly, the kinetics of oxygen electro-reduction are affected by methanol crossover. The mechanism of this impact, which decreases the pace of response, remains unclear to this day. Although increasing the amount of platinum-ruthenium catalyst has a negative effect on cost, choosing suitable membranes and oxygen-tolerant cathodes may decrease reaction rates.

In most cases, fuel crossover will automatically lower fuel consumption, impact and decrease cathode activity, produce excess heat, and impair cell potential and aggregate performance. The primary elements that lead to fuel crossover issues are operational circumstances, including temperature, pressure, and flow rate. Thus, research must establish the sweet spot for DMFC concentrations to provide the best results. Lost power from methanol fuel crossover, heat dissipation, and irretrievable electrode reactions causes a 30% reduction in system efficiency in the DMFC system. A decrease in the cathode reaction is caused by the high methanol crossover, which, in turn, reduces the efficacy of the cathode reduction process. An improved membrane and catalyst are required for this purpose but are also somewhat expensive. Maximizing system efficiency is achieved by reducing fuel permeability.

Most DLFC types address fuel crossover problems, which significantly affect overall performance. Low fuel concentration and membrane permeability are two issues that have prompted several attempts to address them. Nevertheless, an additional consequence of the low fuel concentration is that DMFC performance will decline with decreasing amounts of fuel molecules to oxidize.

The present research presents new nanocomposite polymer membranes for DMFCs that include surface-modified sepiolite (MS) into a PVDF-g-PS copolymer as the PEM. Various methods are used to examine the effects of chemically altering the surface of sepiolite (SP) using vinyltriethoxysilane. Sulfonic acid (SO_3H) active sites on the surface of nanocomposite PVDF-g-PS/MS membranes are created using the phase inversion approach and then treated with chlorosulfonic acid. The water absorption percentage, cation exchange capacity, proton conductivity (σ), and methanol permeability are the physicochemical properties that are examined in the produced nanocomposite membranes (S-PPMS). Because of the surface functional -SO3H groups and well-dispersed MS, the proton conductivity of the S-PPMS membranes is enhanced. At 110°C, S-PPMS10 achieves a peak power density of 210 mW cm^{-2}, surpassing the output achieved by Nafion-117. According to these encouraging findings, the developed nanocomposite PEMs show promise for DMFC application.

One other way to get around fuel crossover is to make better catalysts. To avoid fuel loss and overpotential, these catalysts exhibit inert properties of any reaction that shouldn't occur at the cathode, improving the system's overall efficiency. The catalyst's inertness in the cathode ensures that the reduction may proceed uninterrupted.

The cathode-side efficiency and excellent inertness towards fuel oxidation of transition metal-based catalysts make them a possible platinum replacement. A drawback of non-platinum catalysts is their poor ORR activity capacity, necessitating using additives or supporting materials to improve their effectiveness. Support materials improve ORR performance because of their large surface area, enhancing the catalyst's reactive site count. The cathode side of DMFC devices is treated with oxygen gas, an oxidizing agent. Researchers also like hydrogen peroxide as an oxidizing agent since it is more efficient than oxygen due to its faster reaction rate. On the other hand, breakdown issues mean that hydrogen peroxide can't function in electron systems based on Pt, Pd, or Au. Water is the byproduct of cathode reduction, independent of the agent type. Cathode flooding and a drop in electroreduction activity result from excess water content, which blocks off some of the catalyst's reactive sites and generally lowers performance. Transforming the GDL into a water filter using hydrophobic filters is one approach to the problem. This hydrophobic layer optimizes water management and boosts DMFC system efficiency by acting as a waterproof layer on the cathode side.

Concerns about the cell's stability during long-term operation have slowed the commercialization of DMFCs. Since DMFCs are poisoned by the intermediates formed when methanol is oxidized to MEA, their degradation rates are likely to be significant. Based on crystal size and catalyst agglomeration, *Cheng et al.* observed that cell performance degraded significantly after 200 hours of operation and worsened after 1,002 hours. The original performances of DMFCs could not be regained when their performances were reduced to varying degrees, according to *Wang et al.* The PtRu black catalyst's performance may have degraded due, in part, to the dissolution of Ru metal from anodic catalyst surfaces. Similar findings were reported by *Sarma et al.*, who discovered that after 600 hours, the cathode-exposed membrane degrades more severely than the anode-exposed membrane due to the dissolution of Ru metal. The overall performance of the catalyst is affected by ruthenium, which, once deposited on the cathode, hinders oxygen reduction kinetics and the catalyst's capacity to tolerate methanol crossover. Finding a better DMFC catalyst, such as a binary or ternary catalyst, is necessary to address this difficulty.

Initially developed for PEMFCs, DuPont's Nafion is now the de facto standard for DMFC membranes. With its excellent ionic conductivity and chemical stability, the Nafion membrane is comparatively long-lasting, with an unmatched lifetime of > 60,000 h in the PEMFC. Nevertheless, Nafion membranes do not work well in DMFCs due to a number of drawbacks, such as a high crossover of methanol and ruthenium (in PtRu anodes), a high price tag (\$600–\$1,200 m^{-2}), a low-temperature limit (< 80°C), and excessive humidification. All of these things point to the fact that DMFCs need improved membranes. Lower cost, broader temperature range (80–180°C), more excellent ionic conductivity, and less methanol and ruthenium crossover were some of the benefits identified for composite fluorinated and nonfluorinated (hydrocarbon) DMFC membranes compared to Nafion membranes. While acidic-basic composites (e.g., polyaryl) and inorganic chemicals (SiO_2, silanes, Zr, etc.) enhance Nafion-based membranes, they do not lower cost. For DMFCs, hydrocarbon membranes outperform Nafion membranes in cost and technical efficacy. In addition to improved stability and conductivity, they exhibit reduced methanol crossover.

Before systems, including DMFC, can be marketed effectively, it is crucial to verify their durability, one of the most critical performance factors. A high-quality system can continue to operate for an extended amount of time without degrading performance from when it was first installed. The DMFC system's performance will be impacted as the reaction time grows. System instability occurs due to catalyst accumulation, changes in electrode porosity, and deterioration of electrolyte membranes. The PtRu catalyst is used as the anode catalyst in almost all DMFC systems. As the catalyst dissolves, the anode catalyst's potential drops, which helps Ru cross across to the electrolyte and end up on the cathode. Consequently, the cathode process will be less efficient, and oxygen reduction will not occur.

Due to inadequate oxygen flow into the cathode layer, poor water management, and cathode flooding caused by degrading catalysts, the hydrophobic characteristics of the cathode may diminish. Because of this issue, the system's performance will gradually decrease. *Cheng et al.* attested that after 200 hours of testing, the cell's performance degraded; after 1,002 hours, the overall operation degraded.

A primary objective for every company is to attain commercialization, production, and energy consumption cost competitiveness of DMFC. Ultimately, the cost will determine the price that the end customer pays. The need for a large quantity of platinum catalysts to get an appropriate power density is one of the main obstacles for DMFCs to achieve commercialization. Two methods were used to lower the price of the fuel cell catalysts: first, to search for non-noble catalysts, and second, to decrease the Pt loading. Several proposals for DMFCs have used non-noble catalysts, such as heat-treated macrocycles, transition metal oxides, transition metal chalcogenides, and so on. But their activity levels aren't high enough to be considered platinum substitutes.

Modern technology has made great strides in reducing costs and increasing the DMFC system's usable life. Many research industries have tested the endurance of their conceptual DMFC devices for more than 5,000 hours. However, details like the pricing of the devices are not yet available. Getting a better understanding of the cost savings offered by massive industrial manufacturing has received less attention than recent efforts to reduce the price of individual components while increasing performance. The financial gains needed to boost demand adoption may be substantial if these fuel cells were mass-produced at a rate proportional to their commercial output. Membranes, noble metal catalysts, electrodes, bipolar plates, clamping plates, current collectors, gasket pieces, peripherals, assembly procedures, and overall cost are the main factors that determine the price of a fuel cell stack. An estimate of labor, electricity, and capital expenses must also be included. The modular fuel cell system makes it perfect for integrated mass production. In order to achieve maximum competitiveness in industrial use, output, and energy consumption, the producer's primary objective is to minimize the cost of the DMFC. As governments and industries worldwide seek to reduce carbon emissions and transition to cleaner energy sources, DMFCs are gaining attention for their environmental benefits. Policies promoting renewable energy and low-emission technologies are expected to drive demand for DMFCs. Ongoing research and development efforts are leading to significant improvements in DMFC performance, durability, and cost-effectiveness. Innovations in catalyst materials, membrane technology, and system integration are enhancing the commercial viability of DMFCs, opening up new market opportunities. The growth of renewable methanol production, driven by advances in biomass gasification, carbon capture and utilization, and electrochemical reduction, is expected to support the adoption of DMFCs. As renewable methanol becomes more widely available, the carbon footprint of DMFCs will be further reduced, making them more attractive for various applications.

3.6 Future Perspectives and Research Directions

As the world moves towards sustainable and clean energy solutions, DMFCs are poised to play a significant role. This section explores the future trends in DMFC technology, critical areas for further research and development, and the potential role of DMFCs in the future energy landscape. The future of DMFC technology is shaped by ongoing advancements aimed at improving performance, reducing costs, and expanding applications. Key trends include the following:

Because of their extended runtimes, DMFCs' durability is a significant issue. Among the problems that might arise in DMFC, methanol crossover, mass transport losses, ohmic losses, anodic and cathodic crossover losses, and activation overpotential rank high among the most important ones. Because of this, starting with a single-cell DMFC and optimizing its performance before building a DMFC stack is recommended to boost the power output. It is advised that the impact of the MEA, bipolar plate, and flow field design of a single-cell DMFC be investigated. To get a high-power density, the MEA must be adjusted, and the methanol crossover must be reduced, which requires using the MEA at high methanol concentrations. The flow field design must minimize anode and cathode flooding for gases like CO_2 and water. Because of its role as a current collector, the bipolar plate material is also crucial. In order to create the stack DMFC, the ideal single-cell design is required. With this layout, the stack DMFC may function at peak efficiency. An active mechanism is used in the stack DMFC design to improve the performance of DMFCs at high power output while ensuring their stability and longevity. For the DMFC to function at its best, a comprehensive active system study using the software is advised. Following this, the stack arrangement may be used to apply the active system and achieve the DMFC's optimal performance.

High energy conversion efficiency, mobility, absence of toxic pollutants, and ease of transportation and storage make DMFCs attractive power sources. New catalyst developments have opened doors to commercial DMFC applications by dramatically increasing MOR activity and durability. A lot of work remains, but DMFCs have a lot of promise. Performance drops when the transition metals dissolve in the fuel cell's operation. Secondly, the thick catalyst layer in the MEA affects the gas diffusion and proton transfer, which reduces utilization efficiency because the transition metals can't oxidize methanol very well. Therefore, a substantial amount of the alloyed electrocatalyst is required to get the typical Pt loading (2–3 mg cm^{-2}) on the carbon paper.

The future perspectives and research directions for DMFCs are promising, with ongoing advancements poised to address current challenges and unlock new applications. By focusing on catalyst development, membrane innovation, system optimization, and renewable methanol production, researchers and industry stakeholders can enhance the performance, durability, and economic viability of DMFCs. As part of a broader transition to sustainable energy, DMFCs have the potential to play a crucial role in various sectors, from transportation to decentralized power generation. Continued investment in research and development, along with supportive policies and market incentives, will be essential for realizing the full potential of DMFC technology and contributing to a cleaner, more sustainable energy future.

The future of DMFCs is shaped by ongoing research and development aimed at improving performance, reducing costs, and expanding applications. Critical areas for further research include advanced catalyst development, innovative membrane materials, system optimization, and renewable methanol production. As part of a broader transition to sustainable energy, DMFCs have the potential to contribute significantly to a cleaner and more resilient energy system. DMFCs are poised to play a crucial role in the global energy transition, offering a sustainable and efficient alternative to fossil fuels. Their versatility and adaptability make them suitable for a wide range of applications, from portable devices to large-scale power generation. By leveraging renewable methanol production and integrating with other clean energy technologies, DMFCs can help reduce greenhouse gas emissions and enhance energy security. The continued development and deployment of DMFC technology will require collaboration between researchers, industry stakeholders, and policymakers. Investment in research and development, supportive regulatory frameworks, and market incentives are essential for overcoming the challenges and realizing the full potential of DMFCs. As technological advancements continue to drive improvements in performance and cost-effectiveness, DMFCs are expected to become an integral part of the sustainable energy landscape. DMFCs represent a promising and versatile technology with the potential to address some of the most pressing challenges in the energy sector. Their environmental benefits, economic viability, and adaptability to various applications make them valuable to the portfolio of clean energy solutions. As the world moves towards a more sustainable and resilient energy future, DMFCs are poised to play a key role in this transition, contributing to a cleaner, greener, and more sustainable world. By focusing on technological advancements, addressing challenges, and leveraging market opportunities, DMFCs can unlock new possibilities in the energy landscape. The journey towards widespread adoption and commercialization of DMFC technology is ongoing, but the progress made so far is promising. With continued innovation and collaboration, DMFCs have the potential to significantly impact the way we generate and consume energy, paving the way for a sustainable future.

3.7 Conclusion

The current state of DMFC technology, along with its environmental and policy implications, have all been thoroughly investigated in this research. DMFC technology does away with the complicated procedures involved in converting methanol into hydrogen by directly converting the chemical energy contained in liquid methanol to electrical energy. Methanol fuel has a number of advantages over compressed hydrogen (at 1000 bar) and even liquid hydrogen, including a lower price per unit of energy and a far greater volumetric and gravimetric energy density. Modern Li-ion batteries have a specific energy density that is ten times lower than that of methanol fuel. The DMFC has gained interest from the transportation and portable electronics industries for the earlier reasons. Compared to batteries, the DMFCs are now undergoing fast development and offer the best hope of replacing them as the future power source for portable applications. Its fast functioning, low complexity, and excellent energy density are the primary reasons for this. Nevertheless, a number of significant challenges must be overcome before DMFC can be broadly used. Cell performance is significantly reduced due to the most outstanding technical hurdle: methanol crossover across the membrane. Cell operating parameters, methanol input concentration, and membrane structure and thickness are all variables. Reducing fuel crossover is as simple as using a diluted methanol solution and a high current density. The commercialization of DMFC is also influenced by its stability, affordability, and durability. It will be quite some time before DMFC is commercially available due to the high quantity of platinum catalysts needed at both electrodes to improve electrochemical activity. In order to develop a trustworthy DMFC technology that can be used for many purposes, more studies into these key issues are required.

DMFCs represent a promising technology in the realm of sustainable energy solutions. As the world grapples with the challenges of climate change, energy security, and the need for clean energy sources, DMFCs offer a viable alternative to traditional fossil fuel–based power generation. This chapter has explored the fundamental principles, technological advancements, performance metrics, environmental and economic benefits, challenges and limitations, applications, market potential, and future perspectives of DMFCs. DMFCs directly convert chemical energy from methanol into electrical energy through electrochemical reactions. Key components include the anode, cathode, electrolyte, and membrane. The direct use of liquid methanol as a fuel offers advantages in storage and transport compared to hydrogen, making DMFCs suitable for a variety of applications.

Significant technological advancements have been made in DMFCs, particularly in catalyst development, membrane materials, and system design. Innovations in catalysts, such as using non-precious metals and nanostructured materials, have enhanced performance and reduced costs. Improvements in membrane technology, including composite and hybrid membranes, have addressed methanol crossover and durability issues. Optimized cell designs, such as microfluidic systems and integrated hybrid configurations, have further improved efficiency and reliability. The performance of DMFCs is measured using metrics such as voltage efficiency, fuel utilization efficiency, overall energy efficiency, and power density. Various factors influence these metrics, including catalyst activity, membrane properties, and operating conditions. High-performance DMFC systems have demonstrated their effectiveness in portable electronics, backup power systems, and military applications. DMFCs offer significant environmental benefits, including reduced greenhouse gas emissions and lower pollution levels than traditional energy sources. The potential for renewable methanol production further enhances their sustainability. Economically, ongoing technological advancements and reductions in component costs are making DMFCs more competitive with other energy solutions. The analysis of lifecycle costs and market trends indicates promising economic viability for DMFC technology.

Despite their advantages, DMFCs face several challenges, including methanol crossover, catalyst degradation, membrane durability, and water management. Economic and scalability issues, such as high costs and production challenges, also need to be addressed. Safety concerns related to methanol toxicity and flammability require robust mitigation strategies to ensure safe operation.

DMFCs have a wide range of applications, from portable electronics and backup power systems to transportation and stationary power generation. Emerging applications and market trends highlight the growing demand for clean energy solutions and the potential for DMFCs to play a significant role. Integration with renewable energy systems and advancements in renewable methanol production are expected to drive further market growth.

References

[1] D. Divan, S. Sharma, The Next 20 Years—Utilities Get Ready for Decarbonization, *Energy*. 2040 (2024) 113–138. https://doi.org/10.1007/978-3-031-49417-8_5.

[2] G. Upreti, Climate Change and Its Threat to Humanity in the Anthropocene, *Ecosociocentrism*. (2023) 137–162. https://doi.org/10.1007/978-3-031-41754-2_7.

[3] G. MacDonald, Climate, Capital, Conflict: Geographies of Success or Failure in the Twenty-First Century, *Ann. Am. Assoc. Geogr.* 110 (2020) 2011–2031. https://doi.org/10.1080/24694452.2020.1800300.
[4] H. Mei, Y.P. Li, C. Suo, Y. Ma, J. Lv, Analyzing the Impact of Climate Change on Energy-Economy-Carbon Nexus System in China, *Appl. Energy.* 262 (2020) 114568. https://doi.org/10.1016/J.APENERGY.2020.114568.
[5] R. Kumar, The United Nations and Global Environmental Governance, *Strateg. Anal.* 44 (2020) 479–489. https://doi.org/10.1080/09700161.2020.1824462.
[6] I. Stoddard, K. Anderson, S. Capstick, W. Carton, J. Depledge, K. Facer, C. Gough, F. Hache, C. Hoolohan, M. Hultman, N. Hällström, S. Kartha, S. Klinsky, M. Kuchler, E. Lövbrand, N. Nasiritousi, P. Newell, G.P. Peters, Y. Sokona, A. Stirling, M. Stilwell, C.L. Spash, M. Williams, Three Decades of Climate Mitigation: Why Haven't We Bent the Global Emissions Curve? *Annu. Rev. Environ. Resour.* 46 (2021) 653–689. https://doi.org/10.1146/ANNUREV-ENVIRON-012220-011104/CITE/REFWORKS.
[7] A.A. Akaev, O.I. Davydova, The Paris Agreement on Climate Is Coming into Force: Will the Great Energy Transition Take Place? *Her. Russ. Acad. Sci.* 90 (2020) 588–599. https://doi.org/10.1134/S1019331620050111/FIGURES/6.
[8] Y. Ghorbani, S.E. Zhang, G.T. Nwaila, J.E. Bourdeau, D.H. Rose, Embracing a Diverse Approach to a Globally Inclusive Green Energy Transition: Moving Beyond Decarbonisation and Recognising Realistic Carbon Reduction Strategies, *J. Clean. Prod.* 434 (2024) 140414. https://doi.org/10.1016/J.JCLEPRO.2023.140414.
[9] L. Pang, L. Liu, X. Zhou, M. Hafeez, S. Ullah, M.T. Sohail, How Does Natural Resource Depletion Affect Energy Security Risk? New Insights from Major Energy-Consuming Countries, *Energy Strateg. Rev.* 54 (2024) 101460. https://doi.org/10.1016/J.ESR.2024.101460.
[10] J. Wang, W. Azam, Natural Resource Scarcity, Fossil Fuel Energy Consumption, and Total Greenhouse Gas Emissions in Top Emitting Countries, *Geosci. Front.* 15 (2024) 101757. https://doi.org/10.1016/J.GSF.2023.101757.
[11] K. Hooshyari, B. Amini Horri, H. Abdoli, M. Fallah Vostakola, P. Kakavand, P. Salarizadeh, A Review of Recent Developments and Advanced Applications of High-Temperature Polymer Electrolyte Membranes for PEM Fuel Cells, *Energies.* 14 (2021) 5440. https://doi.org/10.3390/EN14175440.
[12] P. Pei, M. Wang, D. Chen, P. Ren, L. Zhang, Key Technologies for Polymer Electrolyte Membrane Fuel Cell Systems Fueled Impure Hydrogen, *Prog. Nat. Sci. Mater. Int.* 30 (2020) 751–763. https://doi.org/10.1016/J.PNSC.2020.08.015.
[13] Y. Wang, H. Yuan, A. Martinez, P. Hong, H. Xu, F.R. Bockmiller, Polymer Electrolyte Membrane Fuel Cell and Hydrogen Station Networks for Automobiles: Status, Technology, and Perspectives, *Adv. Appl. Energy.* 2 (2021) 100011. https://doi.org/10.1016/J.ADAPEN.2021.100011.
[14] R.M. Nauman Javed, A. Al-Othman, M. Tawalbeh, A.G. Olabi, Recent Developments in Graphene and Graphene Oxide Materials for Polymer Electrolyte Membrane Fuel Cells Applications, *Renew. Sustain. Energy Rev.* 168 (2022) 112836. https://doi.org/10.1016/J.RSER.2022.112836.
[15] B.C. Tashie-Lewis, S.G. Nnabuife, Hydrogen Production, Distribution, Storage and Power Conversion in a Hydrogen Economy—A Technology Review, *Chem. Eng. J. Adv.* 8 (2021) 100172. https://doi.org/10.1016/J.CEJA.2021.100172.
[16] A.B. Avagyan, B. Singh, Biodiesel: Feedstocks, Technologies, Economics and Barriers, Biodiesel Feed. *Technol. Econ. Barriers* (2019). https://doi.org/10.1007/978-981-13-5746-6.

[17] O. Muraza, Highlighting the Greener Shift in Transportation Energy and Fuels Based on Novel Catalytic Materials, *Energy Fuels.* 35 (2021) 25–44. https://doi.org/10.1021/ACS.ENERGYFUELS.0C03105/ASSET/IMAGES/MEDIUM/EF0C03105_0009.GIF.

[18] A. Saeed, G. Shabir, A. Khurshid, Biofuel Molecules, *Recent Adv. Ind. Biochem.* (2024) 147–224. https://doi.org/10.1007/978-3-031-50989-6_9.

[19] Y. Du, Z. Guo, H. Bao, Smooth Sailing Ahead? Policy Options for China's New Energy Vehicle Industry in the Post-Subsidy Era, *Energy Res. Soc. Sci.* 107 (2024) 103359. https://doi.org/10.1016/J.ERSS.2023.103359.

[20] X. Ren, T.E. Springer, S. Gottesfeld, Water and Methanol Uptakes in Nafion Membranes and Membrane Effects on Direct Methanol Cell Performance, *J. Electrochem. Soc.* 147 (2000) 92. https://doi.org/10.1149/1.1393161/XML.

[21] S. Petrovic, E. Hossain, Development of a Novel Technological Readiness Assessment Tool for Fuel Cell Technology, *IEEE Access.* 8 (2020) 132237–132252. https://doi.org/10.1109/ACCESS.2020.3009193.

[22] G. Shen, J. Liu, H. Bin Wu, P. Xu, F. Liu, C. Tongsh, K. Jiao, J. Li, M. Liu, M. Cai, J.P. Lemmon, G. Soloveichik, H. Li, J. Zhu, Y. Lu, Multi-Functional Anodes Boost the Transient Power and Durability of Proton Exchange Membrane Fuel Cells, *Nat. Commun.* 11 (2020) 1–10. https://doi.org/10.1038/s41467-020-14822-y.

[23] Y. Sun, L. Cui, J. Gong, J. Zhang, Y. Xiang, S. Lu, Design of a Catalytic Layer with Hierarchical Proton Transport Structure: The Role of Nafion Nanofiber, *ACS Sustain. Chem. Eng.* 7 (2019) 2955–2963. https://doi.org/10.1021/ACSSUSCHEMENG.8B03910/SUPPL_FILE/SC8B03910_SI_001.PDF.

[24] P.C. Okonkwo, I. Ben Belgacem, W. Emori, P.C. Uzoma, Nafion Degradation Mechanisms in Proton Exchange Membrane Fuel Cell (PEMFC) System: A Review, *Int. J. Hydrogen Energy.* 46 (2021) 27956–27973. https://doi.org/10.1016/J.IJHYDENE.2021.06.032.

[25] D. Akinyele, E. Olabode, A. Amole, Review of Fuel Cell Technologies and Applications for Sustainable Microgrid Systems, *Inventions.* 5 (2020) 42. https://doi.org/10.3390/INVENTIONS5030042.

[26] M.H. de Sá, A.M.F.R. Pinto, V.B. Oliveira, Passive Small Direct Alcohol Fuel Cells for Low-Power Portable Applications: Assessment Based on Innovative Increments Since 2018, *Energies.* 15 (2022) 3787. https://doi.org/10.3390/EN15103787.

[27] C. Houchins, G.J. Kleen, J.S. Spendelow, J. Kopasz, D. Peterson, N.L. Garland, D.L. Ho, J. Marcinkoski, K.E. Martin, R. Tyler, D.C. Papageorgopoulos, U.S. DOE Progress Towards Developing Low-Cost, High Performance, Durable Polymer Electrolyte Membranes for Fuel Cell Applications, *Membranes.* 2 (2012) 855–878. https://doi.org/10.3390/MEMBRANES2040855.

[28] N.A.A. Qasem, G.A.Q. Abdulrahman, A Recent Comprehensive Review of Fuel Cells: History, Types, and Applications, *Int. J. Energy Res.* 2024 (2024) 7271748. https://doi.org/10.1155/2024/7271748.

[29] N. Shaari, S.K. Kamarudin, R. Bahru, S.H. Osman, N.A.I. Md Ishak, Progress and Challenges: Review for Direct Liquid Fuel Cell, *Int. J. Energy Res.* 45 (2021) 6644–6688. https://doi.org/10.1002/ER.6353.

[30] M.M. Tellez-Cruz, J. Escorihuela, O. Solorza-Feria, V. Compañ, Proton Exchange Membrane Fuel Cells (PEMFCs): Advances and Challenges, *Polymers.* 13 (2021) 3064. https://doi.org/10.3390/POLYM13183064.

[31] J. Wang, System Integration, Durability and Reliability of Fuel Cells: Challenges and Solutions, *Appl. Energy.* 189 (2017) 460–479. https://doi.org/10.1016/J.APENERGY.2016.12.083.

[32] S.S. Pethaiah, K.K. Sadasivuni, A. Jayakumar, D. Ponnamma, C.S. Tiwary, G. Sasikumar, Methanol Electrolysis for Hydrogen Production Using Polymer Electrolyte Membrane: A Mini-Review, *Energies.* 13 (2020) 5879. https://doi.org/10.3390/EN13225879.

[33] W. Sun, W. Zhang, H. Su, P. Leung, L. Xing, L. Xu, C. Yang, Q. Xu, Improving Cell Performance and Alleviating Performance Degradation by Constructing a Novel Structure of Membrane Electrode Assembly (MEA) of DMFCs, *Int. J. Hydrogen Energy.* 44 (2019) 32231–32239. https://doi.org/10.1016/J.IJHYDENE.2019.10.113.

[34] S. Jung, High-Performance Direct Methanol Fuel Cell with Optimized Membrane Electrode Assembly: A Theoretical Approach, *Int. J. Hydrogen Energy.* 41 (2016) 9022–9034. https://doi.org/10.1016/J.IJHYDENE.2016.03.147.

[35] H. Burhan, K. Cellat, G. Yılmaz, F. Şen, Direct Methanol Fuel Cells (DMFCs), *Direct Liq. Fuel Cells Fundam. Adv. Futur* (2021) 71–94. https://doi.org/10.1016/B978-0-12-818624-4.00003-0.

[36] E. Giordano, E. Berretti, L. Capozzoli, A. Lavacchi, M. Muhyuddin, C. Santoro, I. Gatto, A. Zaffora, M. Santamaria, Boosting DMFC Power Output by Adding Sulfuric Acid as a Supporting Electrolyte: Effect on Cell Performance Equipped with Platinum and Platinum Group Metal-Free Cathodes, *J. Power Sources.* 563 (2023) 232806. https://doi.org/10.1016/J.JPOWSOUR.2023.232806.

[37] X. Li, Z. Miao, L. Marten, I. Blankenau, Experimental Measurements of Fuel and Water Crossover in an Active DMFC, *Int. J. Hydrogen Energy.* 46 (2021) 4437–4446. https://doi.org/10.1016/J.IJHYDENE.2020.11.027.

[38] A. Jayakumar, D.K. Madheswaran, N.M. Kumar, A Critical Assessment on Functional Attributes and Degradation Mechanism of Membrane Electrode Assembly Components in Direct Methanol Fuel Cells, *Sustain.* 13 (2021) 13938. https://doi.org/10.3390/SU132413938.

[39] D.S. Dhawale, S. Biswas, G. Kaur, S. Giddey, Challenges and Advancement in Direct Ammonia Solid Oxide Fuel Cells: A Review, *Inorg. Chem. Front.* 10 (2023) 6176–6192. https://doi.org/10.1039/D3QI01557B.

[40] A.T. Hamada, M.F. Orhan, A.M. Kannan, Alkaline Fuel Cells: Status and Prospects, *Energy Rep.* 9 (2023) 6396–6418. https://doi.org/10.1016/J.EGYR.2023.05.276.

[41] W.W. Ng, H.S. Thiam, Y.L. Pang, K.C. Chong, S.O. Lai, A State-of-Art on the Development of Nafion-Based Membrane for Performance Improvement in Direct Methanol Fuel Cells, *Membranes.* 12 (2022) 506. https://doi.org/10.3390/MEMBRANES12050506.

[42] M.H. de Sá, A.M.F.R. Pinto, V.B. Oliveira, Passive Direct Methanol Fuel Cells as a Sustainable Alternative to Batteries in Hearing Aid Devices—An Overview, *Int. J. Hydrogen Energy.* 47 (2022) 16552–16567. https://doi.org/10.1016/J.IJHYDENE.2022.03.146.

[43] X. Chen, Z. Zhang, J. Shen, Z. Hu, Micro Direct Methanol Fuel Cell: Functional Components, Supplies Management, Packaging Technology and Application, *Int. J. Energy Res.* 41 (2017) 613–627. https://doi.org/10.1002/ER.3634.

[44] M.S. Alias, S.K. Kamarudin, A.M. Zainoodin, M.S. Masdar, Active Direct Methanol Fuel Cell: An Overview, *Int. J. Hydrogen Energy.* 45 (2020) 19620–19641. https://doi.org/10.1016/J.IJHYDENE.2020.04.202.

[45] A. Sajid, E. Pervaiz, H. Ali, T. Noor, M.M. Baig, A Perspective on Development of Fuel Cell Materials: Electrodes and Electrolyte, *Int. J. Energy Res.* 46 (2022) 6953–6988. https://doi.org/10.1002/ER.7635.
[46] M.I. Awad, W.S. Al-Saidi, M.S. El-Deab, B.A. Al-Jahdaly, M.A. Kassem, Enhanced Electro-Oxidation of Methanol at Nanoparticle-Based Ru/Pt Bimetallic Catalyst: Impact of GC Substrate Pretreatment, *Int. J. Hydrogen Energy.* 45 (2020) 27171–27181. https://doi.org/10.1016/J.IJHYDENE.2020.07.056.
[47] S. Yin, K. Zong, L. Ban, H. Yin, X. Liu, Y. Ding, Design of Compressively Strained PtRu Alloy as Anode for High-Performance DMFCs, *Inorg. Chem. Front.* 11 (2024) 3633–3642. https://doi.org/10.1039/D4QI00478G.
[48] M. Haskul, A.T. Ülgen, A. Döner, Fabrication and Characterization of Ni Modified TiO2 Electrode as Anode Material for Direct Methanol Fuel Cell, *Int. J. Hydrogen Energy.* 45 (2020) 4860–4874. https://doi.org/10.1016/J.IJHYDENE.2019.12.060.
[49] Z. Cai, M. Kamiko, I. Yamada, S. Yagi, PtCo3Nanoparticle-Encapsulated Carbon Nanotubes as Active Catalysts for Methanol Fuel Cell Anodes, *ACS Appl. Nano Mater.* 4 (2021) 1445–1454. https://doi.org/10.1021/ACSANM.0C02977/SUPPL_FILE/AN0C02977_SI_001.PDF.
[50] P. Mukherjee, B. Kakade, A. Swami, Current Trends in Platinum-Based Ternary Alloys as Promising Electrocatalysts for the Oxygen Reduction Reaction: A Mini Review, *Energy Fuels.* 36 (2022) 2306–2322. https://doi.org/10.1021/ACS.ENERGYFUELS.1C03667/ASSET/IMAGES/MEDIUM/EF1C03667_0009.GIF.
[51] D. Lee, S. Gok, Y. Kim, Y.E. Sung, E. Lee, J.H. Jang, J.Y. Hwang, O.J. Kwon, T. Lim, Methanol Tolerant Pt-C Core-Shell Cathode Catalyst for Direct Methanol Fuel Cells, *ACS Appl. Mater. Interfaces.* 12 (2020) 44588–44596. https://doi.org/10.1021/ACSAMI.0C07812/SUPPL_FILE/AM0C07812_SI_001.PDF.
[52] C. Hou, W. Yuan, Y. Zhang, Y. Zhang, X. Zhang, A Hybrid Cathode Catalytic Layer Composed of M–N–C and Pt/C for Direct Methanol Fuel Cells with High Methanol Tolerance, *Fuel.* 368 (2024) 131553. https://doi.org/10.1016/J.FUEL.2024.131553.
[53] Y. Fang, T. Zhang, Y. Zhang, J. Zhu, Evolution of Oxygen Reduction Property by the Role of CeO2 in CeO2/MnO2 Cathode Catalyst for Passive Direct Methanol Fuel Cells, *Appl. Catal. A Gen.* 627 (2021) 118378. https://doi.org/10.1016/J.APCATA.2021.118378.
[54] N. Seselj, S.M. Alfaro, E. Bompolaki, L.N. Cleemann, T. Torres, K. Azizi, Catalyst Development for High-Temperature Polymer Electrolyte Membrane Fuel Cell (HT-PEMFC) Applications, *Adv. Mater.* 35 (2023) 2302207. https://doi.org/10.1002/ADMA.202302207.
[55] J. Hou, M. Yang, C. Ke, G. Wei, C. Priest, Z. Qiao, G. Wu, J. Zhang, Platinum-group-Metal Catalysts for Proton Exchange Membrane Fuel Cells: From Catalyst Design to Electrode Structure Optimization, *EnergyChem.* 2 (2020) 100023. https://doi.org/10.1016/J.ENCHEM.2019.100023.
[56] C.Y. Ahn, J.E. Park, S. Kim, O.H. Kim, W. Hwang, M. Her, S.Y. Kang, S. Park, O.J. Kwon, H.S. Park, Y.H. Cho, Y.E. Sung, Differences in the Electrochemical Performance of Pt-Based Catalysts Used for Polymer Electrolyte Membrane Fuel Cells in Liquid Half-and Full-Cells, *Chem. Rev.* 121 (2021). https://doi.org/10.1021/ACS.CHEMREV.0C01337/ASSET/IMAGES/MEDIUM/CR0C01337_0041.GIF.
[57] M. Chen, C. Zhao, F. Sun, J. Fan, H. Li, H. Wang, Research Progress of Catalyst Layer and Interlayer Interface Structures in Membrane Electrode Assembly (MEA) for Proton Exchange Membrane Fuel Cell (PEMFC) System, *ETransportation.* 5 (2020) 100075. https://doi.org/10.1016/J.ETRAN.2020.100075.

[58] N. Niluroutu, K. Pichaimuthu, S. Sarmah, P. Dhanasekaran, A. Shukla, S.M. Unni, S.D. Bhat, A Copper–Trimesic Acid Metal–Organic Framework Incorporated Sulfonated Poly(ether Ether Ketone) Based Polymer Electrolyte Membrane for Direct Methanol Fuel Cells, *New J. Chem.* 42 (2018) 16758–16765. https://doi.org/10.1039/C8NJ03459A.

[59] F. Zhong, Z. Zeng, Y. Liu, R. Hou, X. Nie, Y. Jia, J. Xi, H. Liu, W. Niu, F. Zhang, Modification of Sulfonated Poly (Etherether Ketone) Composite Polymer Electrolyte Membranes with 2D Molybdenum Disulfide Nanosheet-Coated Carbon Nanotubes for Direct Methanol Fuel Cell Application, *Polymer (Guildf).* 249 (2022) 124839. https://doi.org/10.1016/J.POLYMER.2022.124839.

[60] V. Raj, Direct Methanol Fuel Cells in Portable Applications: Materials, Designs, Operating Parameters, and Practical Steps Toward Commercialization, *Direct Methanol Fuel Cell Technol.* (2020) 495–525. https://doi.org/10.1016/B978-0-12-819158-3.00016-1.

[61] A. Kumar, P. Daw, D. Milstein, Homogeneous Catalysis for Sustainable Energy: Hydrogen and Methanol Economies, Fuels from Biomass, and Related Topics, *Chem. Rev.* 122 (2022) 385–441. https://doi.org/10.1021/ACS.CHEMREV.1C00412/ASSET/IMAGES/MEDIUM/CR1C00412_0046.GIF.

[62] S.S. Araya, V. Liso, X. Cui, N. Li, J. Zhu, S.L. Sahlin, S.H. Jensen, M.P. Nielsen, S.K. Kær, A Review of The Methanol Economy: The Fuel Cell Route, *Energies.* 13 (2020) 596. https://doi.org/10.3390/EN13030596.

[63] M. Romagnoli, V. Testa, M. Romagnoli, V. Testa, Perspective Chapter: Methanol as a Fuel for Direct Methanol Fuel Cells (DMFCs)—Principles and Performance, *Methanol Fuel Transp. Sect. Fuel Cells.* (2023). https://doi.org/10.5772/INTECHOPEN.1002872.

[64] P. Giridhar, S.R.K. Neeraja, Air Pollution a Major Health Hazard In Future, *Asian J. Pharm. Res. Dev.* 8 (2020) 118–121. https://doi.org/10.22270/AJPRD.V8I3.729.

[65] I. Manisalidis, E. Stavropoulou, A. Stavropoulos, E. Bezirtzoglou, Environmental and Health Impacts of Air Pollution: A Review, *Front. Public Heal.* 8 (2020) 505570. https://doi.org/10.3389/FPUBH.2020.00014/BIBTEX.

[66] J. Ren, Proton Exchange Membrane Fuel Cells (PEMFC) and Energy Storage System (ESS) Passive Hybridization for Electric Vehicles (2021). https://pure.coventry.ac.uk/ws/portalfiles/portal/43079418/Ren2021.pdf (accessed August 31, 2024).

[67] C. Dall'Armi, D. Pivetta, R. Taccani, Hybrid PEM Fuel Cell Power Plants Fuelled by Hydrogen for Improving Sustainability in Shipping: State of the Art and Review on Active Projects, *Energies.* 16 (2023) 2022. https://doi.org/10.3390/EN16042022/S1.

[68] B.P. Sandaka, J. Kumar, Alternative Vehicular Fuels for Environmental Decarbonization: A Critical Review of Challenges in Using Electricity, Hydrogen, and Biofuels as a Sustainable Vehicular Fuel, *Chem. Eng. J. Adv.* 14 (2023) 100442. https://doi.org/10.1016/J.CEJA.2022.100442.

4

Direct Ethanol Fuel Cells: Engineering Towards Net-Zero Carbon Emission

Naeemakhtar Momin and Jayappa Manjanna

4.1 Introduction

The increasing global demand for energy, coupled with a rapidly expanding population, has resulted in escalated utilization of fossil fuels. This surge has led to an increase in CO_2 emissions and other greenhouse gases, which are key contributors to global warming and various environmental issues [1–3]. To counteract these negative effects, there is a pressing need to bolster the use of renewable energy sources and their associated distributed conversion technologies worldwide, especially in developing regions [4,5]. Among the various eco-friendly alternative energy conversion devices, fuel cells stand out due to their ability to directly convert chemical energy from fuel into electricity through electrochemical reactions [6]. Fuel cells are recognized for providing an eco-friendly alternative to conventional fossil fuel engines and are regarded as a promising technology for the future of the clean energy sector [7,8]. In particular, DEFCs have captured significant interest because of their high efficiency and environmentally friendly characteristics. With their theoretical high energy density, sustainability, portability, storability, and unique physicochemical properties, DEFCs are considered highly competitive among commercial fuel cells [9]. However, the typical DEFC design, which uses acidic proton exchange membranes (PEMs), often experiences inefficient ethanol oxidation, leading to compromised energy efficiency [10–12].

Recent advancements suggest that replacing acidic electrolytes with alkaline alternatives, such as hydroxide exchange ionomer and hydroxide exchange membranes, can significantly improve ethanol electro-oxidation kinetics, thereby enhancing overall cell performance [13]. Progress in alkaline environments has also positively impacted the oxygen reduction reaction (ORR) and the ethanol electro-oxidation reaction (EOR). The use of more

DOI: 10.1201/9781003512592-4

affordable non-metal electrocatalysts further positions alkaline DEFCs as a promising option [14,15]. Direct methanol fuel cells (DMFCs), which utilize methanol, a readily oxidizable simple alcohol with a methyl group connected to a hydroxyl group, have also attracted significant attention. However, DMFCs face various technical issues, such as severe methanol crossover, slow anode reaction kinetics, and safety concerns due to methanol's toxicity and volatility, which hinders their large-scale deployment [16]. Ethanol, by contrast, is less toxic and offers a more energy-efficient alternative with the potential for sustainable production from agricultural byproducts or biomass [17]. Regarded as a carbon-neutral fuel over its lifecycle, ethanol serves as a renewable energy source with many advantages, including ease of handling, transport, storage, and sustainable production, making it an appealing candidate for a fuel supply. Consequently, DECs hold substantial potential as an alternative energy source for portable applications, thanks to their higher energy density [18–20].

The main objective of DEFCs is to oxidize ethanol to carbon dioxide, a process that releases 12 electrons per ethanol molecule, compared to the six electrons from DMFC oxidation. However, DEFCs often exhibit lower oxidation rates, influenced by variables such as catalyst type, membrane quality, fuel composition, and operational conditions, like temperature and concentration. Thus, ongoing research endeavors aim to improve the energy efficiency of ethanol use in DEFCs (Figure 4.1) [21–23].

Fuel cells are known for their superior fuel conversion and electrical efficiencies compared to conventional combustion or power generation methods, producing minimal toxic emissions, such as NOx, ozone, and particulate matter [24]. With advantages like modularity, quiet operation, high fuel utilization rates, combined heat and power generation capabilities at the point of use, and reduced greenhouse gas emissions, fuel cells present as an enticing solution for future transportation, stationary, and portable power needs, offering significant long-term economic and environmental benefits [25]. A major barrier to the broader commercialization of fuel cell technologies, particularly in the transportation sector, is the current challenge of establishing cost-effective hydrogen storage methods and the nearly absent hydrogen transport and distribution infrastructure [26–28]. The utilization of liquid fuels, such as ethanol and methanol in fuel cells, can bypass the need for extensive infrastructure modifications required for hydrogen-based systems [29]. Notably, infrastructure for using ethanol as a transport fuel is already in place or can be readily adapted. Brazil is a prime example of bioethanol production and use in transport, where bioethanol is in notable use [30].

Biofuels like methanol and ethanol present a CO_2-neutral power generation option when derived from biomass, contributing to an overall reduction in CO_2 emissions and other pollutants [31]. With the prospect of reaching peak oil production and declining petroleum reserves, biofuels also offer regional security for transport fuel supplies [32]. The raw materials for bioethanol production are diverse and include sugarcane, corn, beetroot, wheat,

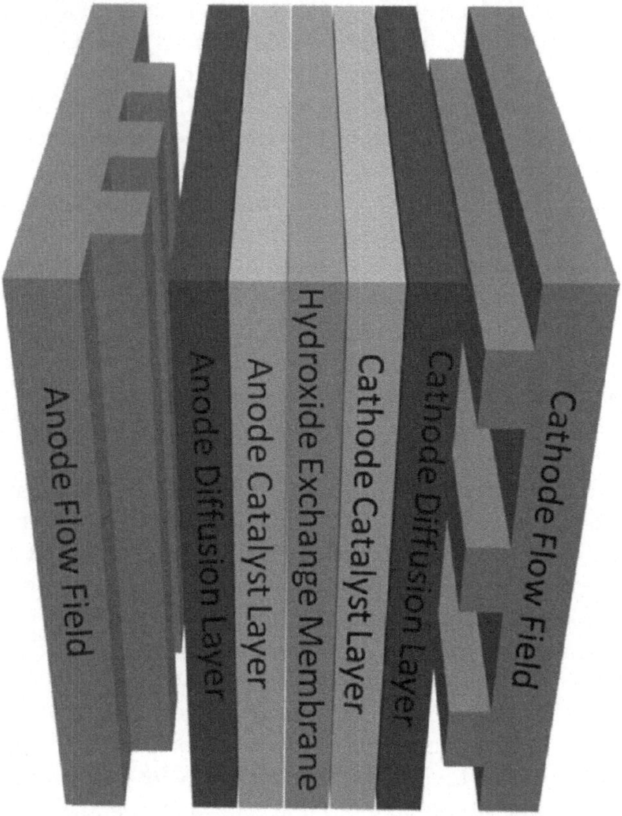

FIGURE 4.1
A typical alkaline direct ethanol fuel cell (DEFC) [21].

Source: With permission from Elsevier. Copyright© 2024.

soybean, or even abundant low-grade cellulosic materials, like woodchips and agricultural or municipal waste [33,34]. The United States and Brazil are the leading producers, accounting for roughly 87% of global ethanol output [35]. Although the competition between bioethanol production from food crops and the food supply chain raises concerns, the energy balance is generally favorable. Producing ethanol from non-edible biomass has become an increasingly favored alternative [36]. The energy return on investment for bioethanol, particularly from sugarcane, is usually higher than that from corn [37].

In the context of alcohol-based fuel cells, significant development efforts are underway for systems utilizing direct (where methanol is consumed directly in the electrochemical reactions) and indirect methanol systems (converting methanol to hydrogen for traditional fuel cells), particularly for portable power generation. Liquid fuel-based fuel cells currently present

TABLE 4.1

Properties and Conversion Efficiencies of Different Alcohols Being Considered as Fuels for Fuel Cells and Their Comparison with Hydrogen [73]*

Property	Hydrogen	Methanol	Ethanol	Ethylene Glycol
Formula	H_2	CH_3OH	C_2H_5OH	$(CH_2OH)_2$
$-\Delta G^o$ (kJ/mol)	237	702	1325	1180
$-\Delta H$ (kJ/mol)	286	726	1367	1192
Energy density, LHV (kW h/kg)	33	6.09	8.00	5.29
Energy density, LHV (kW h/L)	2.96×10^{-3}	4.80	6.32	5.80
E^o_{cell} (V)	1.23	1.21	1.14	1.22
n_{exp}	2	4	4	8
n_{theo}	2	6	12	10
Energy stored (Ah/kg/Ah/L)	26,802/2.40	3350/2653	2330/1841	3458/3855
ε_{cell}^{rev} (%)	83	97	97	99
ε_f (%)	100	67	33	88
$^b\varepsilon_V$ (%)	57	41	44	41
ε_{cell} ($\varepsilon_{cell}^{rev} \times \varepsilon_f \times \varepsilon_V$)	54	27	14	36
		40[a]	43[a]	41[a]

*Adapted from Badwal et al. With permission from Elsevier. Copyright© 2024.
E^o_{cell}: Reversible cell voltage.
ε_{cell}^{rev}: Theoretical electric efficiency (or reversible cell efficiency).
ε_f: Faradic or current efficiency.
ε_V: Voltage efficiency of the cell.
ε_{cell}: The overall efficiency of the cell.
n_{actual}: Actual number of electrons involved in the oxidation reaction.
n_{theo}: Theoretical number of electrons involved in the complete fuel oxidation reaction.
[a] In case complete electrochemical reaction occurs with the involvement of all electrons.
[b] Assuming cell operating voltage of 0.7 V for hydrogen and 0.5 V for alcohols [73].

complex balance-of-plant requirements due to the necessary upfront fuel processing. While methanol's high toxicity has pushed research towards ethanol being renewable, widely available, affordable, and non-toxic, its application as a fuel cell resource is in need of development. The power densities and lifetimes of ethanol fuel cells fall short of the requirements for commercialization, largely due to slow electrode kinetics, which require 12 electrons per molecule for ethanol's electrochemical oxidation, and the challenge of C-C bond dissociation (Figure 4.2) [31,38].

However, advancements in developing suitable anode catalysts that can facilitate ethanol oxidation and cathode catalysts resistant to ethanol and its byproducts are in progress. Current ethanol-fuelled polymer electrolyte membrane systems necessitate high loadings of precious metal electrocatalysts, unlike traditional hydrogen-based fuel cells. Operating at higher

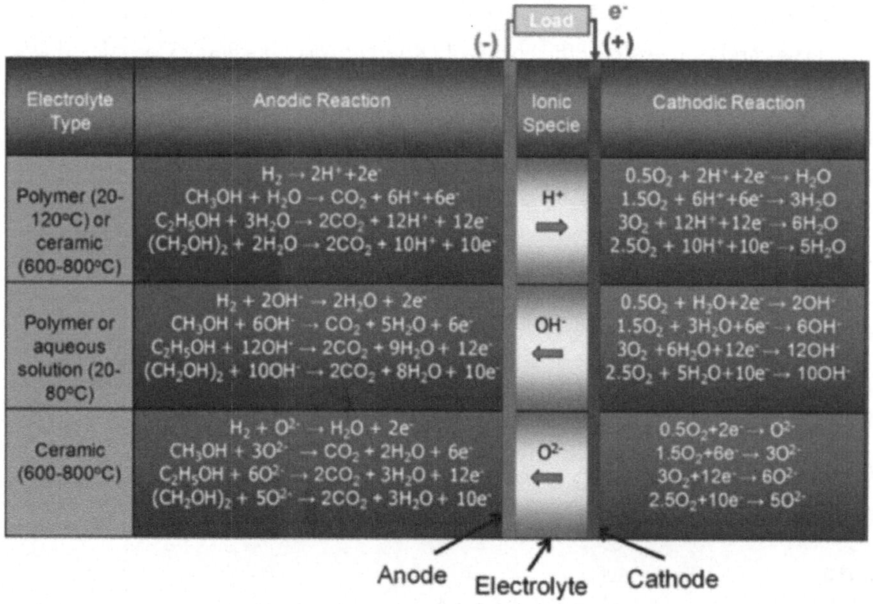

FIGURE 4.2
Schematic showing electrochemical reactions involved in various types of fuel-based alcohol fuel cells [73].

Source: With permission from Elsevier. Copyright© 2024.

temperatures could potentially speed up reaction kinetics, reduce catalyst poisoning, and permit the use of less expensive catalysts [39]. This paper reviews various types of DEFCs, addressing aspects such as ethanol sources, production methods, cell construction materials, operating regimes, cell and stack fabrication, performance, lifetime concerns, technology statuses, and market applications [40].

Direct alcohol fuel cells (DAFCs) are an emerging category of energy converters that have received extensive interest, including alkaline fuel cells (AFCs), which have shown higher current densities than proton exchange membrane fuel cells (PEMFCs). Although PEMFCs have recently witnessed increased interest due to solid electrolyte flexibility and leak prevention, AFCs still offer advantages, such as lower costs and corrosivity [41]. They also enable the use of affordable non-noble metal electrocatalysts, like nickel, silver, and palladium, due to faster oxygen reduction reaction kinetics compared to PEMFCs [42].

DAFCs primarily use alcohols, such as methanol, ethanol, ethylene glycol, and 2-propanol as fuels, boasting energy densities comparable to hydrocarbons and gasoline. Alkaline media DAFCs, in specific, have shown excellent performance. Historically, DAFCs used acid proton exchange membranes and platinum-based catalysts, which facilitated CO_2 removal from anode reactions but hindered overall performance due to suboptimal alcohol electro-oxidation kinetics. Transitioning to alkaline electrolytes or anion-exchange membranes has highlighted better oxidation polarization characteristics for methanol on platinum in alkaline vs. acidic media and reduced methanol permeation rates due to the reversal of ionic current, allowing cheaper metal catalysts to be used [43]. Yet DAFCs encounter challenges, such as poor electrocatalytic performance at lower temperatures and anode surface poisoning by CO-like intermediates [44]. Dual-media DEFCs, combining an alkaline anode and an acidic cathode, are limited by the low theoretical voltage of 1.14 V, which necessitates using a cation exchange membrane/PEM over anion-exchange membranes (AEMs), which have been largely disregarded due to poor thermal stability and lower ionic conductivity than PEMs (Figure 4.3) [45].

Past research predominantly focused on developing DMFCs but was burdened by slow-reaction kinetics for methanol oxidation, methanol crossover

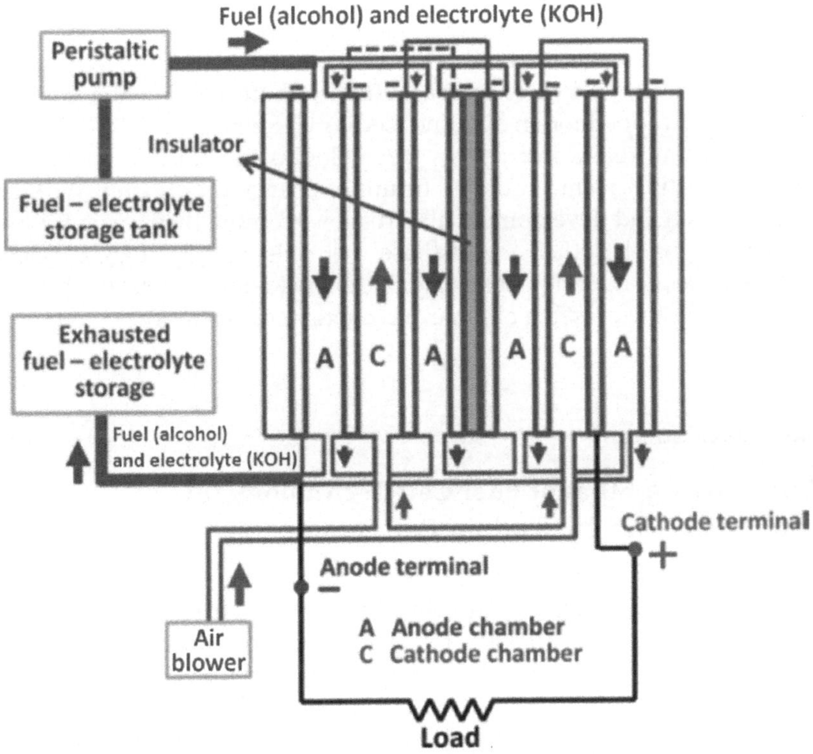

FIGURE 4.3
Schematic of a direct alcohol alkaline four-cell fuel cell stack [73].

Source: With permission from Elsevier. Copyright© 2024.

through Nafion membranes to the oxygen electrode, and anode poisoning by strongly adsorbed intermediates [46]. The methanol crossover leads to a decrease in fuel efficiency and the formation of mixed potentials on the cathode, which lowers the potential and wastes fuel. DMFCs also struggle with CO poisoning of Pt catalysts at low temperatures and the high material costs associated with the membrane, catalyst, and separator. Considering methanol's non-renewability, toxicity, volatility, and flammability, its use in portable devices poses considerable challenges [47].

Ethanol is less toxic and has a higher energy density that can be sustainably produced through agricultural bioprocesses, marking it as a renewable source of energy. Studies indicate that ethanol's crossover rate is lower and has a less detrimental effect on cathode performance compared to methanol. Recent electrocatalyst development has focused on PtRu and PtSn catalysts for DAFCs, with PtRu/C being appropriate for methanol and PtSn/C particularly suited for ethanol in acidic settings [44]. There are two types of direct ethanol fuel cells: proton exchange membrane DEFCs (PEM-DEFCs) and anion-exchange membrane DEFCs (AEM-DEFCs). PEM-DEFCs encounter issues with the slow kinetics of ethanol oxidation reactions (EOR), and despite the use of acidic media and Pt catalysts, these difficulties remain [45]. However, advancements in alkaline media have demonstrated faster kinetics for EOR and ORR, leading to increased fuel cell efficiency when the acidic electrolyte is swapped for an alkaline medium [48,49].

In conclusion, renewable energy technologies like DEFCs and DMFCs present promising avenues for the future of energy production. With continued research and development aimed at overcoming their current limitations, enhancing the efficiency of these fuel cells, and leveraging existing infrastructure, the adoption of sustainable energy sources is set to play a key role in tackling the pressing climate and energy challenges of our era.

4.2 Improving Ethanol Fuel Cell Technology for a Future with No Carbon Emissions

As more people live on Earth and we need more energy, we keep using more oil and gas. This makes problems like global warming worse because of the extra CO_2 in the air [50]. To fix this, we need to use energy from things like the sun and wind more because they don't harm the environment. They turn energy around us into electricity that we can use. Fuel cells are one good way to do this. They don't pollute the air and can change energy from fuels directly into electricity. One kind of fuel cell that is good for the planet is called a DEFC. DEFCs work well because they can hold a lot of energy, are easy to get, and can be moved around safely. Plus, ethanol, the fuel they use, has some helpful features [51]. People like using DEFCs more and more for business. But DEFCs aren't perfect. They don't work as well in acid because it's hard to turn ethanol into electricity. Scientists are trying to make them better by using different materials that work well in a base instead of an acid. This makes the process of turning ethanol into electricity and of using oxygen work better [52]. They're also trying out cheaper materials that don't use expensive metals and are still good at making electricity [53].

Methanol is another alcohol used in fuel cells. It's easier to turn it into electricity, but methanol isn't great because it's poisonous, can evaporate quickly, and can cause technical problems. Ethanol is safer, works better, and we can make it from plants or leftover farm materials without hurting the planet when we get it from the right sources [31,54].

DEFCs converts ethanol into CO_2 and along with the production of electricity. But many things can affect how well ethanol changes into electricity, like what the fuel cell is made of and how hot or concentrated the fuel is. So we need more research to make these fuel cells better [55]. Fuel cells are better than old ways of making power because they turn fuel into energy very well, don't pollute as much, and don't give off as many greenhouse gases [56]. They're also good because you can use them to make heat and power where you need them. This is good for the future of transport and portable power [57–59].

But it's hard to start using fuel cells everywhere, especially for cars, because they can be expensive, don't last very long, and there's no good system to get hydrogen, their fuel, to people. Using liquids like ethanol and methanol in fuel cells is easier because we already know how to get them around. Ethanol is especially good because we already have ways to distribute it [60–62]. In some countries, people use a kind of fuel called bioethanol for their cars. We can make it from different plants, like sugarcane, corn, or leftovers from farms and forests [63]. The United States and Brazil are really good at making ethanol, which helps them use less of other types of fuels. But some people are worried that making biofuels might use up too much energy and might compete with growing food. So using parts of plants that we don't eat for food is a better choice [64–67].

Even with other kinds of alcohol fuel cells being looked into, ethanol seems really promising because it's easy to get, doesn't cost too much, is safe, and is good for the planet [68]. Scientists are working on making better fuel cells with ethanol, both DEFCs and alkaline fuel cells (AFCs). AFCs can make more power than some other fuel cells and are less likely to leak [69]. But AFCs could be cheaper because they use less expensive metals and aren't as harsh on materials. Still, even with great progress, we have obstacles, like fuel cells not working well in the cold and parts being ruined by other chemicals that get in the way [70]. To make better fuel cells, we need new ideas, especially for creating strong parts that don't break down over time and can handle turning ethanol and oxygen into electricity even when it's hot [71].

TABLE 4.2
Different Systems of Alkaline Direct Ethanol Fuel Cells [87]*

Fuel	Anode	Cathode	Electrolyte	Temperature (°C)	MPD (mW cm^{-2})
Methanol + KOH	Pt/C	Pt/C	AHA Tokuyama	50	5.5
EG + KOH					9.2
Methanol + NaOH	Pt/C	Pt/C	Morgane®-ADP, Solvay	60	6.8
Methanol + NaOH	Pt/Ti	Pt/C	Morgane®-ADP, 50Solvay	60	7.8
Methanol + KOH	PtRu/C	Pt/C	AHA Tokuyama	50	6.2
EG + KOH (1 M)		Pt/C			9.5
EG + KOH (3 M)		Ag/C			8.1
Methanol + NaOH	Pt/C	Pt/C	Morgane®-ADP, Solvay	20	18
EG + NaOH					19
EG + NaOH	Pt/C	Pt/C	Morgane®-ADP, Solvay	20	19
	PtPb/C				22
	PtPbPd/C				28
	Pt/C	Pt/C	Morgane®-ADP, Solvay	20	19
		PtPb/C			23.5
EG + KOH	PtRu/C	Pt/C	A-006, Tokuyama	80	25
		Ag/C			20
		LaSrMnO/C			18
Methanol	PtRu	Pt black	QAETFE	50	1.5
				80	8.5
Methanol + KOH	PtRu black	MnO_2/C	PVA/SSA	30	2.38
				60	4.13
Methanol + KOH	PtRu black	MnO_2/C	PVA/TiO_2	25	9.25
Ethanol + KOH					8.0
Methanol + KOH	PtRu black	MnO_2/C	PVA/HAP	25	11.48
Ethanol + KOH	PtRu/C	Pt/C	PBI/KOH	75	49.20
				90	60.95
Methanol + KOH	PtRu/C	Pt/C	PBI/KOH	90	31
Methanol	PtRu	Pt black	AAEM-C	50	1.17
Ethanol					1.71
EG					1.57
Ethanol + KOH	PtRu	Pt black	AAEM, Tokuyama	20	58
Methanol + KOH	PtRu/C	Pt/C	AAEM, Tokuyama	20	6.8

*Adapted from Kamarudin et al. With permission from Elsevier. Copyright© 2024.

TABLE 4.3

Several Models of Alkaline Direct Ethanol Fuel Cell with Maximum Power Density (MPD) of Some Pt-Free [87]*

Fuel	Anode	Cathode	Electrolyte	Temperature (°C)	MPD mW cm^{-2}
Ethanol + KOH	Pd–(Ni–Zn)/C	Fe–Co HYPERMEC™	A-006, Tokuyama	20	58
				80	170
Ethanol + KOH	RuV/C	Co-TMPP/C	PBI	80	90 air
					110 O_2
Ethanol (1 M) + KOH (1 M)	Pd_2Ni_3/C	Fe–Co HYPERMEC™	A201, Tokuyama	60	44
Ethanol (3 M) + KOH (5 M)				60	90
Ethanol + KOH	Ni–Fe–Co HYPERMEC™	Fe–Co HYPERMEC™	A201, Tokuyama	40	60
Methanol + KOH	Pd/MWCNT	Fe–Co HYPERMEC™	A-006, Tokuyama	20–22	7
Ethanol + KOH					18
EG + KOH					5
Ethanol + NaOH	PdNi/C	Pt/C	Nafion 117	60	240
Ethanol + NaOH	PdNi/C	Pt/C	Nafion 211	60	360

*Adapted from Kamarudin et al. With permission from Elsevier. Copyright© 2024.

4.3 The Technical Edge of DEFCs

DEFCs have attracted considerable attention in recent years due to their environmental friendliness and renewable nature. Platinum, a crucial material for the oxidation reaction in fuel cells, can suffer from carbon monoxide poisoning [72,73]. This has led to research into platinum alloys and supporting materials to enhance efficiency without relying solely on pure platinum [74]. Studies have shown that supporting platinum with carbon materials, such as reduced graphene oxide, can significantly enhance oxidation reactions. Similar improvements have been obtained using polyaniline and polypyrrole as supporting materials. Additionally, ongoing research on non-Pt alloys and support materials, such as palladium (Pd), ruthenium-copper (Ru-Cu) alloys, and Pd-cobalt (Co) alloys, has revealed promising catalytic activity [75].

DEFCs are increasingly important in portable devices, including mobile phones and laptops, as well as in the automotive industry [76]. Methanol is often the alcohol of choice due to its low cost, high solubility in electrolytes, and status as the lowest-chain alcohol. However, alcohols like ethanol, ethylene glycol, propanol, and glycerol have also been explored for use in DEFCs [77]. Ethanol is an attractive biofuel, derived from the fermentation of biomass and agricultural products. It is less toxic compared to methanol and exhibits comparable electrochemical activity. Direct ethanol fuel cells are powered by the catalytic oxidation of ethanol, producing water and CO_2 as byproducts in a carbon-neutral cycle. Ethanol offers a higher energy density and lower toxicity [78]. The development of alcohol-based fuel cells includes low-temperature systems (<100°C), high-temperature systems (>400–500°C), as well as proton- and oxygen-ion-conducting electrolytes with liquid or solid bases. Indirect alcohol fuel cells, which convert alcohol to hydrogen and CO before use, also play a vital role in research [79,80].

A major challenge in commercializing DEFCs is the slow electrochemical reactivity at low temperatures, especially due to ethanol's lower specificity for CO_2 over other products, such as acetic acid and acetaldehyde. Pt-based catalysts, while effective at adsorption and dissociation of ethanol, show poor catalytic selectivity, leading to substantial ethanol crossover issues. Development is shifting towards trimetallic catalysts, adding a third metal to PtSn catalysts to improve selectivity and minimize problems, such as CO poisoning [81].

Alkaline DEFCs appear promising due to the potential for developing non-noble catalysts for the ethanol oxidation reaction (EOR) and oxygen reduction reaction (ORR). While platinum displays higher activity in alkaline media, palladium-based catalysts have shown more stability during EOR [82].

Direct ethanol fuel cells are acknowledged as sustainable and carbon-neutral energy sources with numerous benefits for portable and mobile applications, such as higher energy density and ease of transport, storage, and handling compared to hydrogen and methanol [83].

Conventional DEFCs can be divided into proton exchange membrane DEFCs (PEM-DEFCs) and anion-exchange membrane DEFCs (AEM-DEFCs). Although PEM-DEFCs have seen extensive advances, the slow kinetics of EOR in acidic media remains a significant challenge, which might be addressed by using alkaline media. However, to be commercially viable, substantial improvements in cell performance are needed, particularly concerning the physical and chemical properties of the AEM [84]. Moreover, the low theoretical voltage of DEFCs operating in acid and alkaline media is a limitation, potentially surmountable by using hydrogen peroxide (H_2O_2) as an alternative oxidant to increase theoretical voltage and performance [85]. Hydrogen peroxide also enables operation in environments lacking oxygen, such as underwater or in space [86].

In conclusion, DEFC technology holds great promise, yet advancements in catalyst development, membrane materials, and oxidants are essential to achieve the desired efficiency level and enable widespread commercialization.

4.4 Advancements in Ethanol Production and Utilization

4.4.1 Ethanol as a Renewable Energy Source

Addressing the improvement in bioethanol production technologies, this section discusses the diversification of biomass sources and the energy output/input ratios associated with various production methods. It further elaborates on how these advancements substantiate the economic and environmental viability of ethanol as a core element in fuel cells. Renewable energy sources are essential for sustainable development and mitigating global climate concerns. Ethanol, a biofuel, is increasingly in the spotlight as a substitute for fossil fuels. Its production is noteworthy due to its potential to be derived from a variety of biomass sources. These include both food crops, such as sugarcane, corn, and beetroot, and non-food biomass, like switchgrass, poplar, forest residue, and municipal waste [31].

4.4.2 Sources and Energy Output/Input Ratios of Ethanol Production

Ethanol can be produced from the edible parts of plants, as well as inedible plant material and other biomass forms. A critical metric is the energy output/input ratio, which compares the energy content of the ethanol produced to the fossil fuel energy expended in its production, excluding the solar energy the biomass absorbs. For example, corn-derived ethanol can have a ratio of about 2 when produced in facilities powered by fossil fuels. When biomass-generated power covers half of a plant's energy requirement, this ratio can improve to 2.8. Sugarcane presents a more favorable energy balance, with the ratio reaching or surpassing 8 [87].

Further analyses suggest even more substantial ratios for ethanol from cellulosic feedstocks, such as corn stover and sugarcane bagasse. These ratios can range from 5.2 to as much as 32, considering the utilization of residues and co-products in generating electricity or heat for the process [88]. In the United States and Brazil, commercial ethanol production from corn and sugarcane is well-established. Conversely, plants using non-food crop biomass are usually in the pilot or demonstration stage, relying on governmental subsidies and incentives [89].

4.5 Ethanol Production Methods

Ethanol production mainly happens through fermenting starches and sugars from crops, such as corn and sugarcane using yeast. Biomass waste can also be converted using bacteria, or through chemical reactions involving ethene and steam over a catalyst. Over 95% of ethanol is derived from fermentation methods [90].

Bioethanol production processes, for starchy (like corn) and sugary (like sugarcane) raw materials, involve pre-treatment, fermentation, distillation, and separation. Lignocellulosic biomass, with its complex composition of cellulose, hemicellulose, and lignin, requires elaborate pre-treatments, such as hydrolysis and physicochemical procedures, including steam explosion, and enzymatic steps to break down cellulose for fermentation. Innovations, such as incorporating specialized yeast enzymes, have been instrumental in increasing ethanol yield [91].

4.5.1 Process Efficiencies and GHG Emission Reduction

Assessing efficiency often involves looking at the ratio of the energy content of ethanol to that of the input biomass. Sugarcane and corn fermentation processes can exceed 40% energy efficiency, while advanced methods for lignocellulosic substrates can reach around 50% efficiency for the ethanol, with additional energy generated from the usage of residues [92]. Reducing greenhouse gas (GHG) emissions is another essential objective of ethanol production. Traditional production with crops like sugarcane and corn can greatly decrease GHG emissions, and advanced methods for lignocellulosic feedstocks promise even further reductions, partly due to the allocation of GHG reduction credits from co-products [93].

4.5.2 Ethanol in Fuel Cells

Ethanol stands out as a preferred fuel alternative over methanol due to its lower toxicity, higher energy density, and renewable nature. It has a lower crossover rate in fuel cells, helping to prevent degradation of cathode performance. Research efforts have focused on developing catalysts, such as PtRu and PtSn, suitable for DAFCs, considering ethanol's effects on cathode performance and the overall efficiency of the fuel cell [94].

The operational efficiency of DEFCs relies on managing challenges associated with fuel crossover. This is the movement of ethanol through the membrane to the cathode side, which can degrade the catalyst and decrease efficiency. Parameters influencing the ethanol crossover rate, such as temperature, current density, and ethanol concentration, need precise management to optimize DEFC performance [95].

Proposed solutions for reducing ethanol crossover include the development of novel membrane materials like sulfonated poly(ether ether ketone) (SPEEK) and methods that involve membranes coated with carbon molecular sieves or composite membranes blending SPEEK with polyimide. These approaches are aimed at refining DEFCs by minimizing the adverse effects of crossover.

Ethanol is a promising and adaptable renewable energy source suitable for various technological applications. Bioethanol production from an array of biomass sources could potentially cover a significant portion of our energy demands while providing environmental benefits. Continual improvements in production processes and fuel cell technology are enhancing efficiency and reducing GHG emissions, nudging ethanol towards a sustainable and eco-friendly energy future. Nonetheless, ongoing research, technological advancements, and government support remain critical for overcoming existing challenges and scaling up production to satisfy the growing global energy needs [96].

4.6 DEFCs: Electrochemical Mechanism and Materials

In the realm of DEFCs, anhydrous or diluted ethanol is directly supplied to the anode chamber of the cell, while air is channeled to the cathode chamber (Figure 4.4). At the anode/electrolyte boundary, ethanol is oxidized to generate protons (in the case of proton-conducting, electrolyte-based DEFCs), carbon dioxide, electrons, and byproducts, like acetaldehyde and acetic acid [97]. These protons traverse the membrane, whereas the electrons circulate through the electrical load. At the cathode, the protons react with oxygen from the air, culminating in the formation of water. The primary electrochemical reactions can be summarized as follows:

$$\text{Anode reaction: } C_2H_5OH + 3H_2O \rightarrow 2CO_2 + 12H^+ + 12e^- \quad (1)$$

$$\text{Cathode reaction: } 3O_2 + 12H^+ + 12e- \rightarrow 6H_2O \quad (2)$$

$$\text{Overall reaction: } C_2H_5OH + 3O_2 \rightarrow 2CO_2 + 3H_2O \quad (3)$$

Complete ethanol oxidation is characterized by a slow kinetic process due to the transfer of 12 electrons, whereas partial oxidation to acetaldehyde or acetic acid requires only two or four electrons, respectively.

Ethanol's energy density is calculated through its lower heat value (LHV), with the specific and volumetric energy densities being represented as follows:

$$\text{Energy density (kWh/kg)} = -\Delta G^\circ/(3600 \times Mm) \quad (4)$$

$$\text{Energy density (kWh/L)} = -\Delta G^\circ \times \rho/(3600 \times Mm) \quad (5)$$

Where ΔG° (−1325 kJ/mol) is the Gibbs free energy change, Mm is the molar mass of ethanol (0.046 kg/mol), and ρ is the density of ethanol (0.79 kg/L). Using these equations, the energy densities of ethanol are determined to be 8.0 kWh/kg and 6.32 kWh/L. The energy stored in ethanol, expressed in

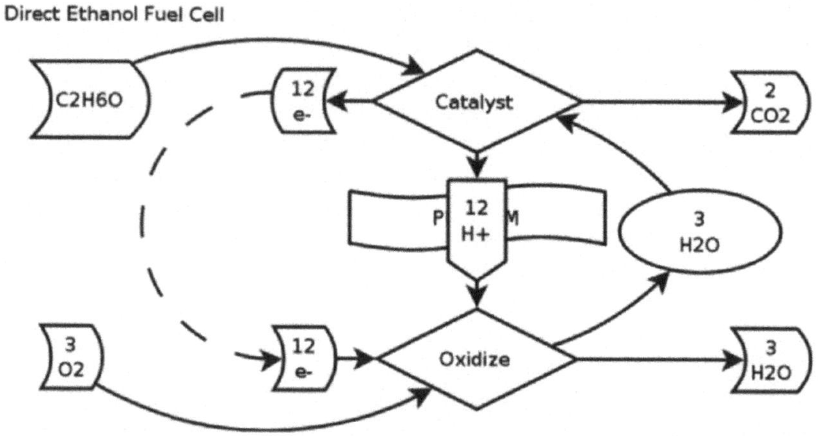

FIGURE 4.4
Reaction flowchart of direct ethanol fuel cell [87].

Source: With permission from Elsevier. Copyright© 2024.

ampere-hours per kilogram (Ah/kg) or per litre (Ah/L), uses the number of electrons transferred in the reaction. The complete oxidation theoretically involves 12-electron transfer, but experiments show the process often stops at the formation of acetic acid, involving only four electrons [98–102].

The theoretical electrical efficiency of ethanol fuel cells is 97%, with the reversible cell voltage calculated to be around 1.14 V assuming complete ethanol oxidation [103]. The efficiency of a DEFC also comprises faradaic efficiency, representing the fraction of electrons contributing to the intended oxidation, and voltage efficiency, which compares the operating cell voltage under load to the reversible voltage. The overall DEFC efficiency for completely oxidized ethanol is estimated around 43% [104].

The research on DEFC technology is primarily intended to overcome the challenges of efficiency, durability, and cost by investigating innovations in catalysts, alternative materials, and fabrication techniques. This includes exploring the potential of alkaline and proton-conducting polymer electrolyte membrane DEFCs and examining anode and cathode catalyst materials, catalyst support systems, polymer electrolyte membrane modifications to address ethanol crossover, and degradation issues to improve long-term durability [105,106].

4.6.1 Technical Challenges and Recent Innovations

Here, the intrinsic technological hurdles encountered in the development of DEFCs are delineated. Progress made in adapting alkaline media for enhanced oxygen reduction and ethanol oxidation kinetics is highlighted, alongside breakthroughs in catalyst development and strategies for mitigating ethanol crossover and poisoning [107]. *High costs of catalyst materials:* Noble metals like platinum (Pt) and platinum-ruthenium (PtRu) are often used as catalysts in DLFCs. These materials are expensive, which raises the cost of fuel cells, making them less competitive against technologies like lithium-ion batteries. Research efforts are exploring ways to reduce, optimize, or replace costly noble catalysts with less expensive materials, like nanostructured titanium nitride (TiN) supports, to improve catalytic activity and stability while reducing costs (Figure 4.5) [108–111].

Sluggish conversion efficiencies: A significant challenge for DLFCs, specifically acidic-type DEFCs, is the low total conversion efficiency of ethanol to carbon dioxide. For instance, methanol crossover in DMFCs leads to only 30% energy exploitation as electricity. This inefficiency is due to various factors including fuel crossover, parasitic reactions, and other limitations of the fuel cells' electrochemical reactions [112–114]. *Limited durability and longevity of cell components:* DLFCs face issues with catalyst deactivation (agglomeration or dissolution), electrolyte degradation, and changes in electrode porosity. Over time, these factors lead to performance degradation. For example, in PtRu catalysts used for the anode, ruthenium may dissolve over time and migrate to the cathode, hindering its efficiency [115,116]. *Technological hurdles in cell and stack fabrication:* These challenges are related to optimizing the overall performance of the fuel cell systems. Efforts are made to create designs that mitigate issues like fuel crossover and improve the utilization of the catalyst by using advanced supporting materials, such as carbon nanofibers [117].

Fuel crossover and cathode flooding: Liquid fuel crossover from the anode to the cathode through the membrane negatively impacts cell performance by reducing fuel utilization and degrading cathode activity. Various strategies are being explored to manage this issue, including optimizing fuel concentrations, developing less permeable membranes, and using hydrophobic air filters [118]. *Cathode catalyst development:* There is a need to develop cathode catalysts that are reactive towards the oxygen reduction reaction (ORR) but inert towards fuel oxidation. This can potentially alleviate the problem of fuel crossover degrading the cathode's performance [119]. *Chemical safety and side product production:* Safety is a concern due to the toxicity and flammability of certain fuels used in DLFCs. Furthermore, the production of poisonous byproducts like carbon monoxide and formaldehyde is an issue needing innovation to reduce poisoning and environmental impact [120]. *Unproven long-term durability:* Long-term durability studies are critical to ensure that DLFCs can maintain their performance over extended periods. Agglomeration of catalyst particles, membrane degradation, and accumulation of poison materials on the anode are identified as contributors to performance decline over time. Research is aimed at enhancing the stability and endurance of DLFC components for better longevity [121,122].

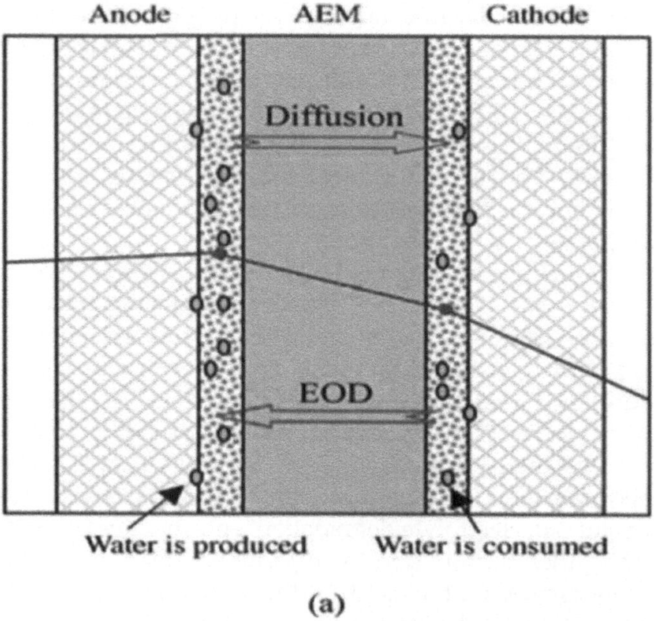

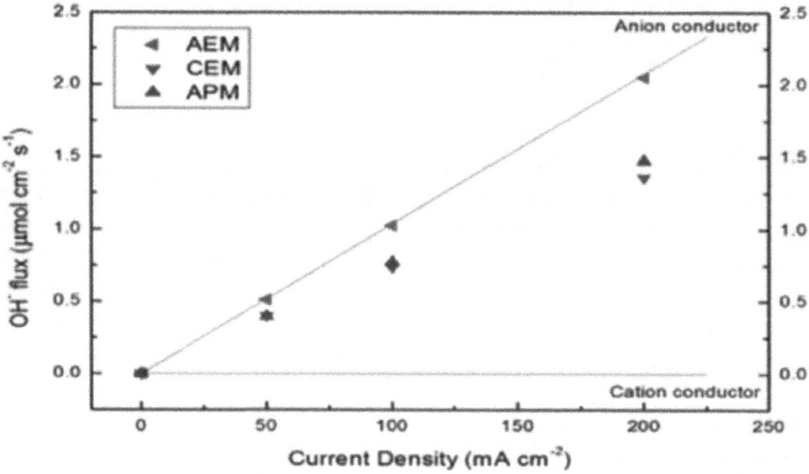

FIGURE 4.5
(a) The typical water distribution through the fuel cell structure in an alkaline DEFC. (b) Variations in the OH flux through three membranes [21].

Source: With permission from Elsevier. Copyright© 2024.

4.7 Challenges

Fuel molecules cross over from the anode to the cathode, which can poison the catalyst and cause undesirable reactions, reducing the cell's efficiency. The kinetics of ethanol oxidation are sluggish at lower temperatures, which hampers the DEFC performance. Accumulation of water in the cathode obstructs oxygen transport, diminishing the fuel cell's performance. The need for platinum-based catalysts inflates the cost, making DEFCs less economically feasible. The DEFC system requires comprehensive modeling to understand the physicochemical behaviors and to optimize performance [123–128]. Heavy reliance on noble metal catalysts, like platinum, increases the cost and loading compared to other technologies, like hydrogen-fed PEMFCs [129]. Fuel crossover occurs when liquid fuel diffuses from the anode to the cathode, leading to reduced efficiency and other issues, such as heat generation and catalyst poisoning [130]. Accumulation of water at the cathode impairs oxygen reduction reaction and overall DLFC performance [131–134].

DLFCs, including DEFCs, offer promise for the future of portable and automotive power; however, a number of challenges pertaining to efficiency, cost, safety, and durability need to be addressed. This would warrant a shift towards designing more robust systems, alternative catalysts, and advanced materials. Moreover, comprehensive system studies and optimization, alongside innovative technical solutions, are quintessential for making these technologies viable for commercialization [135–137].

4.8 Market Applications and Commercial Viability

Shifting the focus to practical applications, this section comprehensively reviews the market penetration of DEFCs in sectors like transportation, portable power, and large-scale stationary power generation. It elucidates the commercialization challenges stemming from system costs, lifespan issues, and infrastructural constraints. Notebooks using DAFCs were first put on the market in 1986. This resulted in 2% notebook use worldwide. In 2008, notebooks provided approximately 150 million buyers all over the world, while in 2009, this ratio reached approximately 180 million [20,138]. Looking directly at alcohol fuel cells, portable devices that work directly with methyl alcohol reached a commercial market of about $190 million in 2020 [70,139]. Initially, the number of notebooks and mobile phones using DAFCs was very low because at that time, procurement of the relevant materials and systems was expensive and time-consuming. Today, with the development of technology, the use of DAFCs has become much more economical in commercial applications and, hence, more prevalent. A portable device that uses

DAFCs is given in Figure 4.5. Methanol is the most studied DAFC, and some companies have been worked directly on the methanol fuel cell. These companies are Direct Methanol Fuel Cell Corporation, Toshiba [140], Panasonic [141], Motorola, Samsung, Fideris (formerly Lynntech Industries), Manhattan Scientifics, NEC, Giner Electrochemical Systems, Yuasa Corporation, MTI Micro Fuel Cells, and Smart Fuel Cell [138]. Motorola was an American telecommunications company founded in 1928; it went out of business in 2011. It has a methanol evaporator and a vapor converter due to its direct alcohol fuel cell system content. Toshiba is a Japanese company selling mobile phones that directly utilize methanol fuel cells. Their devices produce approximately 30 mW/cm^2 of power. In addition, the company announced that it will use fuel cells directly in its laptops, notebooks, and tablets as well. In newly purchased Toshiba devices, the direct alcohol fuel cell needs to be charged for about 12 h before it is more active [140]. The portable chargers manufactured by Toshiba has a 400 mA current and 5 V potential. NEC Corporation is a Japanese commercial firm whose devices have approximately ten times more robust power density than Li-ion based power supplies. Since 2005, it has been trading in notebooks and mobile phones that feature a direct methanol fuel cell system [139].

Yuasa is a Japanese telecommunications firm competing with NEC, Toshiba, and other firms. Its fuel cell devices produce 100–300 W of power [139]. Samsung is a Korean company that produces notebooks that can last 6 h without recharging. In addition, this company plans to reduce its mobile phones with direct methanol fuel cells to the size of automated teller machine (ATM) cards [76]. For portable devices (phones, cameras, etc.), Samsung produced 2.6 W devices with a lifespan of more than 400 hours through replaceable methanol cartridges in 2006, and in these devices, methanol is used as fuel. The devices produced as passive chargers are 5 mm thick and 180 g in weight [141]. Lynntech Industries/Fideris has developed devices with four prototype DAFCs that offer a 15 W/12 V power supply, which it sells to the US Army [139,142–144].

The future outlook of various direct ethanol-based fuel cell systems have been reviewed. Fuel cell commercialization, especially for the transportation and remote area power, is partly hampered by the lack of hydrogen transportation and distribution infrastructure combined with its low volumetric energy density. Ethanol as a fuel is an appealing option for use in fuel cells. In comparison with other fuel cells, direct ethanol fuel cells are very attractive for a number of applications, including transportation, portable power, distributed power, backup and remote area power, and large-scale stationary power generation. The infrastructure for transportation and distribution of ethanol already exists in most countries. The energy density of ethanol is 6.32 kW h/L compared to 0.003 kW h/L for H_2, 4.8 kW h/L for methanol, and 5.80 kW h/L for ethylene glycol. The ethanol also contains 13 wt percent hydrogen. Ethanol can be produced from a range of biomass resources, including sugarcane, corn, grasses, wood and straw. The energy output to input ratio is quite high for sugarcane and many cellulosic materials. Global production of ethanol stands at close to 90 billion litres per annum [145–149].

Ethanol can be utilized directly or indirectly in fuel cells. For indirect ethanol fuel cells, the key technology is up-front ethanol reforming/gas cleaning subsystems (typically steam reforming, followed by shift reaction to convert CO to hydrogen and methanation to remove residual CO). The hydrogen thus produced can be used in most fuel cells. Although the efficiency of such systems is low, they can be used for a much wider range of applications. Ethanol in anhydrous or diluted form (either in liquid or vaporized) can also be directly fed to the anode chamber of the fuel cell and oxidized at the electrode/electrolyte interface offering substantially high efficiency. However, ethanol is a large molecule requiring the transfer of 12 electrons in the electrochemical reaction for its direct oxidation. Therefore, the kinetics of its oxidation is very sluggish, and many intermediate products are formed leading to its incomplete oxidation and poisoning of the catalysts. Often, two or four electron transfer reactions take place in low-temperature DEFCs operating in the vicinity of 100°C or below. Although such systems can be used for the production of both power and chemicals, these lead to Faradaic efficiency loss due to the incomplete conversion of ethanol to CO_2. This also limits the number of applications where DEFCs can be used. Nevertheless, a number of different ethanol fuel cell systems are under development and include both low (< 100°C) and high (> 400–500°C) temperature systems, proton- and oxygen-ion-conducting electrolytes, and liquid or solid electrolyte-based systems [150–154].

The anode catalyst development has been identified as the key technology for most direct ethanol fuel cells in addition to ethanol crossover and poisoning of the catalyst in polymer electrolyte membrane fuel cells and short lifetimes. These issues are less stringent in high-temperature, SOFC-based systems. However, fuel delivery system and carbon deposition on the anode side remain major challenges in these systems. Overall, the technology is still at an early stage of development due to issues discussed earlier with commercialization still further away. Thus far, there have been very few systems demonstrated at the prototype level based on PEM-based fuel cells, either directly fed with ethanol solutions or with hydrogen obtained by ethanol reforming, with sizes ranging from a few watts to the kW range [155–169].

4.9 Future Perspectives

Today, DAFCs are potentially seen to be the most widely used FC types in portable power tools and portable power supplies [170]. Among these,

DEFCs are particularly becoming more promising in portable devices as well as in large-scale applications, such as automobiles. However, acidic-type DEFCs have some limitations due to the low efficiency of PEM membrane DEFCs in total ethanol conversion to carbon dioxide [171]. In general, there are some problems to be solved with the DEFC system. These problems are related to both the immature development of the technical system and economic feasibility. Problems in the system more specifically could be listed as follows: fuel crossover leading to catalyst poisoning and parasitic current formation in the cathode section, decreased ethanol reaction rates at low temperatures, and cathode water flooding preventing oxygen transport. The cost of the catalyst is another problem, which is caused by the need for platinum as the primary catalyst [172–174].

The number of studies on alkaline-acid direct ethanol fuel cells (AA-DEFCs), which perform better between AEM-DEFC and PEM-DEFC types, should be increased. It is particularly remarkable that it provides a power density of 360 mW cm^{-2} at 60°C. In addition, when PEM-DEFC and AEM- DEFC were compared, it was found that PEM-DEFC showed lower performance in studies with both Pt-based catalysts and Pt-free catalysts. To commercialize DEFCs, there is a need to create new isothermal designs and to develop materials suitable for anode fuel channel design. The design efforts are also important to increase the performance of DEFCs. Therefore, in order to develop DEFCs, a multidimensional, multiphase, and dynamic mathematical models are needed urgently to capture the physicochemical behavior of the system. Finally, system optimization, optimal parameters, and total cost study and predictions are also very important for a successful DEFC system implementation [174–179].

4.10 Conclusions

Ethanol, which has high potential in the energy sector as an alternative energy source, is an important energy source due to its portability, low cost, and high energy efficiency. However, because of some unsolved problems in DEFC systems, they still have obstacles to become a commercial product. Hence, there is a need for new studies on DEFCs that will contribute to and develop existing literature knowledge. New generation nanocatalysts based on Pt or Pd should be developed for ethanol oxidation. In order to progress in the commercialization of DEFCs, the development of new designs, in terms of engineering and optimization, are necessary. Furthermore, to develop a better DEFC model, it is necessary to study various situations in more detail, such as fuel management, water management, gas management, multiphase flow, and flow area design.

4.10.1 Acknowledgments

Authors gratefully acknowledge the support from (ii) VGST-CESEM (GRD No. 746), Govt. of Karnataka, India. (iii) DST-FIST, Govt. of India (SR/FST/CSI-273/2016).

4.10.2 Conflicts of Interest

The authors declare that they have no known conflicts of interest/competing financial interests or personal relationships that could have appeared to influence the work reported in this paper.

4.10.3 Credit

Dr. Naeemakhtar Momin: conceptualization, methodology, writing (original draft preparation), data curation, formal analysis, writing (review and editing). *Prof. Jayappa Manjanna:* formal analysis, validation, writing (review and editing).

4.10.4 Declaration of Competing Interest

The authors declare that they have no known competing financial interests or personal relationships that could have appeared to influence the work reported in this paper.

References

[1] S. Chu, A. Majumdar, Opportunities and challenges for a sustainable energy future, *Nature* 488 (2012) 294–303. https://doi.org/10.1038/nature11475.

[2] B. Çelik, G. Başkaya, H. Sert, Ö. Karatepe, E. Erken, F. Şen, Monodisperse Pt(0)/DPA@GO nanoparticles as highly active catalysts for alcohol oxidation and dehydrogenation of DMAB, *Int. J. Hydrogen Energy* 41 (2016) 5661–5669. https://doi.org/10.1016/j.ijhydene.2016.02.061.

[3] S. Ertan, F. Şen, S. Şen, G. Gökağaç, Platinum nanocatalysts prepared with different surfactants for C1–C3 alcohol oxidations and their surface morphologies by AFM, J. Nanoparticle Res. 14 (2012) 922. https://doi.org/10.1007/s11051-012-0922-5.

[4] Z. Ozturk, F. Sen, S. Sen, G. Gokagac, The preparation and characterization of nano-sized Pt–Pd/C catalysts and comparison of their superior catalytic activities for methanol and ethanol oxidation, *J. Mater. Sci.* 47 (2012) 8134–8144. https://doi.org/10.1007/s10853-012-6709-3.

[5] F. Şen, G. Gökağaç, Pt nanoparticles synthesized with new surfactants: Improvement in C1–C3 alcohol oxidation catalytic activity, *J. Appl. Electrochem.* 44 (2014) 199–207. https://doi.org/10.1007/s10800-013-0631-5.

[6] T. Wilberforce, M.A. Abdelkareem, K. Elsaid, A.G. Olabi, E.T. Sayed, Role of carbon-based nanomaterials in improving the performance of microbial fuel cells, *Energy* 240 (2022) 122478. https://doi.org/10.1016/j.energy.2021.122478.

[7] T. Inamuddin, ed., *Microbial Fuel Cells*, Materials Research Forum LLC, 2019. https://doi.org/10.21741/9781644900116.

[8] Mesoporous Materials in Biofuel Cells, 2019, pp. 157–172. https://doi.org/10.21741/9781644900079-7.

[9] H. Burhan, H. Ay, E. Kuyuldar, F. Sen, Monodisperse Pt-Co/GO anodes with varying Pt: Co ratios as highly active and stable electrocatalysts for methanol electrooxidation reaction, *Sci. Rep.* 10 (2020) 6114. https://doi.org/10.1038/s41598-020-63247-6.

[10] M. Winter, R.J. Brodd, What are batteries, fuel cells, and supercapacitors? *Chem. Rev.* 104 (2004) 4245–4270. https://doi.org/10.1021/cr020730k.

[11] E. Antolini, Palladium in fuel cell catalysis, *Energy Environ. Sci.* 2 (2009) 915. https://doi.org/10.1039/b820837a.

[12] T.S. Zhao, Y.S. Li, S.Y. Shen, Anion-exchange membrane direct ethanol fuel cells: Status and perspective, *Front. Energy Power Eng. China* 4 (2010) 443–458. https://doi.org/10.1007/s11708-010-0127-5.

[13] C. Bianchini, P.K. Shen, Palladium-based electrocatalysts for alcohol oxidation in half cells and in direct alcohol fuel cells, *Chem. Rev.* 109 (2009) 4183–4206. https://doi.org/10.1021/cr9000995.

[14] S. Shen, T. Zhao, One-step polyol synthesis of Rh-on-Pd bimetallic nanodendrites and their electrocatalytic properties for ethanoloxidation in alkaline media, *J. Mater. Chem. A* 1 (2013) 906–912. https://doi.org/10.1039/C2TA00725H.

[15] E.H. Yu, X. Wang, U. Krewer, L. Li, K. Scott, Direct oxidation alkaline fuelcells: From materials to systems, *Energy Environ. Sci.* 5 (2012) 5668–5680. https://doi.org/10.1039/C2EE02552C.

[16] V. Rao, Hariyanto, C. Cremers, U. Stimming, Investigation of the ethanol electro-oxidation in alkaline membrane electrode assembly by differential electrochemical mass spectrometry, *Fuel Cells* 7 (2007) 417–423. https://doi.org/10.1002/fuce.200700026.

[17] Y. Wang, S. Zou, W.-B. Cai, Recent advances on electro-oxidation of ethanol on Pt- and pd-based catalysts: From reaction mechanisms to catalytic materials, *Catalysts* 5 (2015) 1507–1534. https://doi.org/10.3390/catal5031507.

[18] E. Kuyuldar, H. Burhan, A. Şavk, B. Güven, C. Özdemir, S. Şahin, A. Khan, F. Şen, Enhanced electrocatalytic activity and durability of PtRu nanoparticles decorated on rGO material for ethanol oxidation reaction, in *Graphene Functionalization Strategies*, 2019, pp. 389–398. https://doi.org/10.1007/978-981-32-9057-0_16.

[19] W.J. Zhou, S.Q. Song, W.Z. Li, Z.H. Zhou, G.Q. Sun, Q. Xin, S. Douvartzides, P. Tsiakaras, Direct ethanol fuel cells based on PtSn anodes: The effect of Sn content on the fuel cell performance, *J. Power Sources* 140 (2005) 50–58. https://doi.org/10.1016/j.jpowsour.2004.08.003.

[20] L. Wang, A. Lavacchi, M. Bevilacqua, M. Bellini, P. Fornasiero, J. Filippi, M. Innocenti, A. Marchionni, H.A. Miller, F. Vizza, Energy efficiency of alkaline direct ethanol fuel cells employing nanostructured palladium electrocatalysts, *ChemCatChem* 7 (2015) 2214–2221. https://doi.org/10.1002/cctc.201500189.

[21] L. An, T.S. Zhao, Transport phenomena in alkaline direct ethanol fuel cells for sustainable energy production, *J. Power Sources* 341 (2017) 199–211. https://doi.org/10.1016/j.jpowsour.2016.11.117.

[22] S. Guo, S. Dong, E. Wang, Pt/Pd bimetallic nanotubes with petal-like surfaces for enhanced catalytic activity and stability towards ethanol electrooxidation, *Energy Environ. Sci.* 3 (2010) 1307. https://doi.org/10.1039/c0ee00053a.

[23] E. Antolini, E.R. Gonzalez, Alkaline direct alcohol fuel cells, *J. Power Sources* 195 (2010) 3431–3450. https://doi.org/10.1016/j.jpowsour.2009.11.145.

[24] E.H. Yu, U. Krewer, K. Scott, Principles and materials aspects of direct alkaline alcohol fuel cells, *Energies* 3 (2010) 1499–1528. https://doi.org/10.3390/en3081499.

[25] L. Zeng, T.S. Zhao, L. An, A high-performance supportless silver nanowire catalyst for anion exchange membrane fuel cells, *J. Mater. Chem. A* 3 (2015) 1410–1416. https://doi.org/10.1039/C4TA05005C.

[26] J.S. Spendelow, A. Wieckowski, Electrocatalysis of oxygen reduction and small alcohol oxidation in alkaline media, *Phys. Chem. Chem. Phys.* 9 (2007) 2654. https://doi.org/10.1039/b703315j.

[27] N.J. Robertson, H.A. Kostalik, T.J. Clark, P.F. Mutolo, H.D. Abruña, G.W. Coates, Tunable high performance cross-linked alkaline anion exchange membranes for fuel cell applications, *J. Am. Chem. Soc.* 132 (2010) 3400–3404. https://doi.org/10.1021/ja908638d.

[28] F. Ksar, L. Ramos, B. Keita, L. Nadjo, P. Beaunier, H. Remita, Bimetallic palladium–gold nanostructures: Application in ethanol oxidation, *Chem. Mater.* 21 (2009) 3677–3683. https://doi.org/10.1021/cm901364w.

[29] L. An, T.S. Zhao, Z.H. Chai, P. Tan, L. Zeng, Mathematical modeling of an anion-exchange membrane water electrolyzer for hydrogen production, *Int. J. Hydrogen Energy* 39 (2014) 19869–19876. https://doi.org/10.1016/j.ijhydene.2014.10.025.

[30] T.S. Zhao, C. Xu, R. Chen, W.W. Yang, Mass transport phenomena in direct methanol fuel cells, *Prog. Energy Combust. Sci.* 35 (2009) 275–292. https://doi.org/10.1016/j.pecs.2009.01.001.

[31] M.Z.F. Kamarudin, S.K. Kamarudin, M.S. Masdar, W.R.W. Daud, Review: Direct ethanol fuel cells, *Int. J. Hydrogen Energy* 38 (2013) 9438–9453. https://doi.org/10.1016/j.ijhydene.2012.07.059.

[32] M. Ahmed, I. Dincer, A review on methanol crossover in direct methanol fuel cells: Challenges and achievements, *Int. J. Energy Res.* 35 (2011) 1213–1228. https://doi.org/10.1002/er.1889.

[33] X. Zhao, M. Yin, L. Ma, L. Liang, C. Liu, J. Liao, T. Lu, W. Xing, Recent advances in catalysts for direct methanol fuel cells, *Energy Environ. Sci.* 4 (2011) 2736. https://doi.org/10.1039/c1ee01307f.

[34] J.P. Pereira, D.S. Falcão, V.B. Oliveira, A.M.F.R. Pinto, Performance of a passive direct ethanol fuel cell, *J. Power Sources* 256 (2014) 14–19. https://doi.org/10.1016/j.jpowsour.2013.12.036.

[35] S. Song, P. Tsiakaras, Recent progress in direct ethanol proton exchange membrane fuel cells (DE-PEMFCs), *Appl. Catal. B Environ.* 63 (2006) 187–193. https://doi.org/10.1016/j.apcatb.2005.09.018.

[36] G. Andreadis, P. Tsiakaras, Ethanol crossover and direct ethanol PEM fuel cell performance modeling and experimental validation, *Chem. Eng. Sci.* 61 (2006) 7497–7508. https://doi.org/10.1016/j.ces.2006.08.028.

[37] G.M. Andreadis, A.K.M. Podias, P.E. Tsiakaras, A model-based parametric analysis of a direct ethanol polymer electrolyte membrane fuel cell performance, *J. Power Sources* 194 (2009) 397–407. https://doi.org/10.1016/j.jpowsour.2009.04.064.

[38] K.S. Roelofs, T. Hirth, T. Schiestel, Dihydrogenimidazole modified silica-sulfonated poly(ether ether ketone) hybrid materials as electrolyte membranes for direct ethanol fuel cells, *Mater. Sci. Eng. B* 176 (2011) 727–735. https://doi.org/10.1016/j.mseb.2011.02.029.

[39] L. An, T.S. Zhao, R. Chen, Q.X. Wu, A novel direct ethanol fuel cell with high power density, *J. Power Sources* 196 (2011) 6219–6222. https://doi.org/10.1016/j.jpowsour.2011.03.040.

[40] Y.S. Li, T.S. Zhao, A high-performance integrated electrode for anion-exchange membrane direct ethanol fuel cells, *Int. J. Hydrogen Energy* 36 (2011) 7707–7713. https://doi.org/10.1016/j.ijhydene.2011.03.090.

[41] Y.S. Li, T.S. Zhao, R. Chen, Cathode flooding behaviour in alkaline direct ethanol fuel cells, *J. Power Sources* 196 (2011) 133–139. https://doi.org/10.1016/j.jpowsour.2010.06.111.

[42] V. Alzate, K. Fatih, H. Wang, Effect of operating parameters and anode diffusion layer on the direct ethanol fuel cell performance, *J. Power Sources* 196 (2011) 10625–10631. https://doi.org/10.1016/j.jpowsour.2011.08.080.

[43] S. Song, W. Zhou, Z. Zhou, L. Jiang, G. Sun, Q. Xin, V. Leontidis, S. Kontou, P. Tsiakaras, Direct ethanol PEM fuel cells: The case of platinum based anodes, *Int. J. Hydrogen Energy* 30 (2005) 995–1001. https://doi.org/10.1016/j.ijhydene.2004.11.006.

[44] L. An, T.S. Zhao, J.B. Xu, A bi-functional cathode structure for alkaline-acid direct ethanol fuel cells, *Int. J. Hydrogen Energy* 36 (2011) 13089–13095. https://doi.org/10.1016/j.ijhydene.2011.07.025.

[45] S.C. Zignani, V. Baglio, J.J. Linares, G. Monforte, E.R. Gonzalez, A.S. Aricò, Endurance study of a solid polymer electrolyte direct ethanol fuel cell based on a Pt–Sn anode catalyst, *Int. J. Hydrogen Energy* 38 (2013) 11576–11582. https://doi.org/10.1016/j.ijhydene.2013.04.162.

[46] S. Beyhan, C. Coutanceau, J.-M. Léger, T.W. Napporn, F. Kadırgan, Promising anode candidates for direct ethanol fuel cell: Carbon supported PtSn-based trimetallic catalysts prepared by Bönnemann method, *Int. J. Hydrogen Energy* 38 (2013) 6830–6841. https://doi.org/10.1016/j.ijhydene.2013.03.058.

[47] B. Braunchweig, D. Hibbitts, M. Neurock, A. Wieckowski, electrocatalysis: A direct alcohol fuel cell and surface science perspective, *Catal. Today* 202 (2013) 197–209. https://doi.org/10.1016/j.cattod.2012.08.013.

[48] E. Antolini, Catalysts for direct ethanol fuel cells, *J. Power Sources* 170 (2007) 1–12. https://doi.org/10.1016/j.jpowsour.2007.04.009.

[49] H. Burhan, M. Yılmaz, K. Cellat, A. Zeytun, G. Yılmaz, F. Şen, Direct ethanol fuel cells (DEFCs), in *Direct Liq. Fuel Cells*, Elsevier, 2021, pp. 95–113. https://doi.org/10.1016/B978-0-12-818624-4.00004-2.

[50] L. Carrette, K.A. Friedrich, U. Stimming, Fuel cells—Fundamentals and applications, *Fuel Cells* 1 (2001) 5–39. https://doi.org/10.1002/1615–6854(200105)1:1<5::AID-FUCE5>3.0.CO;2-G.

[51] H. Hellman, R. Vandenhoed, Characterising fuel cell technology: Challenges of the commercialisation process, *Int. J. Hydrogen Energy* 32 (2007) 305–315. https://doi.org/10.1016/j.ijhydene.2006.07.029.

[52] S. Giddey, S.P.S. Badwal, A. Kulkarni, C. Munnings, A comprehensive review of direct carbon fuel cell technology, *Prog. Energy Combust. Sci.* 38 (2012) 360–399. https://doi.org/10.1016/j.pecs.2012.01.003.

[53] Ó.J. Sánchez, C.A. Cardona, Trends in biotechnological production of fuel ethanol from different feedstocks, *Bioresour. Technol.* 99 (2008) 5270–5295. https://doi.org/10.1016/j.biortech.2007.11.013.
[54] A. Demirbas, Biofuels securing the planet's future energy needs, *Energy Convers. Manag.* 50 (2009) 2239–2249. https://doi.org/10.1016/j.enconman.2009.05.010.
[55] Y. Yano, D. Blandford, Y.R. Surry, Y. Yano, D. Blandford, Y.R. Surry, From ethanol shuffle to ethanol tourism—why the RFS does not make sense, *Choices Mag. Food, Farm, Resour.* 27 (2012).
[56] M.Q. Wang, GREET 1.5: Transportation fuel-cycle model. Vol. 1: Methodology, development, use, and results (No. ANL/ESD-39-VOL-1). Argonne National Laboratory, 1999.
[57] J. Goldemberg, The ethanol program in Brazil, *Environ. Res. Lett.* 1 (2006) 014008. https://doi.org/10.1088/1748–9326/1/1/014008.
[58] A. Limayem, S.C. Ricke, Lignocellulosic biomass for bioethanol production: Current perspectives, potential issues and future prospects, *Prog. Energy Combust. Sci.* 38 (2012) 449–467. https://doi.org/10.1016/j.pecs.2012.03.002.
[59] C.A. Cardona, Ó.J. Sánchez, Fuel ethanol production: Process design trends and integration opportunities, *Bioresour. Technol.* 98 (2007) 2415–2457. https://doi.org/10.1016/j.biortech.2007.01.002.
[60] A. Kundu, J.H. Jang, J.H. Gil, C.R. Jung, H.R. Lee, S.-H. Kim, B. Ku, Y.S. Oh, Micro-fuel cells—Current development and applications, *J. Power Sources* 170 (2007) 67–78. https://doi.org/10.1016/j.jpowsour.2007.03.066.
[61] T.S. Zhao, C. Xu, Fuel cells—direct alcohol fuel cells | direct methanol fuel cell: Overview performance and operational conditions, in *Encyclopedia of Electrochemical Power Sources*, Elsevier, 2009, pp. 381–389. https://doi.org/10.1016/B978-044452745-5.00244-6.
[62] T.S. Zhao, C. Xu, Fuel cells—direct alcohol fuel cells | direct methanol fuel cell: Overview performance and operational conditions, in *Encyclopedia of Electrochemical Power Sources*, Elsevier, 2009, pp. 381–389. https://doi.org/10.1016/B978-044452745-5.00244-6.
[63] A. Demirbas, Biofuels securing the planet's future energy needs, *Energy Convers. Manag.* 50 (2009) 2239–2249. https://doi.org/10.1016/j.enconman.2009.05.010.
[64] J.T. Eells, M.M. Henry, P. Summerfelt, M.T.T. Wong-Riley, E.V. Buchmann, M. Kane, N.T. Whelan, H.T. Whelan, Therapeutic photobiomodulation for methanol-induced retinal toxicity, *Proc. Natl. Acad. Sci.* 100 (2003) 3439–3444. https://doi.org/10.1073/pnas.0534746100.
[65] D. Jacobsen, K.E. McMartin, Antidotes for methanol and ethylene glycol poisoning, *J. Toxicol. Clin. Toxicol.* 35 (1997) 127–143. https://doi.org/10.3109/15563659709001182.
[66] P. Tsiakaras, A. Demin, Thermodynamic analysis of a solid oxide fuel cell system fuelled by ethanol, *J. Power Sources* 102 (2001) 210–217. https://doi.org/10.1016/S0378-7753(01)00803-5.
[67] W. Jamsak, S. Assabumrungrat, P.L. Douglas, N. Laosiripojana, R. Suwanwarangkul, S. Charojrochkul, E. Croiset, Performance of ethanol-fuelled solid oxide fuel cells: Proton and oxygen ion conductors, *Chem. Eng. J.* 133 (2007) 187–194. https://doi.org/10.1016/j.cej.2007.03.003.
[68] E. Antolini, Catalysts for direct ethanol fuel cells, *J. Power Sources* 170 (2007) 1–12. https://doi.org/10.1016/j.jpowsour.2007.04.009.

[69] L. An, T.S. Zhao, An alkaline direct ethanol fuel cell with a cation exchange membrane, *Energy Environ. Sci.* 4 (2011) 2213. https://doi.org/10.1039/c1ee00002k.

[70] L. Jiang, G. Sun, Fuel cells—direct alcohol fuel cells | direct ethanol fuel cells, in *Encyclopedia of Electrochemical Power Sources*, Elsevier, 2009, pp. 390–401. https://doi.org/10.1016/B978-044452745-5.00245-8.

[71] Y.S. Li, T.S. Zhao, Z.X. Liang, Performance of alkaline electrolyte-membrane-based direct ethanol fuel cells, *J. Power Sources* 187 (2009) 387–392. https://doi.org/10.1016/j.jpowsour.2008.10.132.

[72] M.Z.F. Kamarudin, S.K. Kamarudin, M.S. Masdar, W.R.W. Daud, Review: Direct ethanol fuel cells, *Int. J. Hydrogen Energy* 38 (2013) 9438–9453. https://doi.org/10.1016/j.ijhydene.2012.07.059.

[73] S.P.S. Badwal, S. Giddey, A. Kulkarni, J. Goel, S. Basu, Direct ethanol fuel cells for transport and stationary applications—A comprehensive review, *Appl. Energy* 145 (2015) 80–103. https://doi.org/10.1016/j.apenergy.2015.02.002.

[74] E. Antolini, E.R. Gonzalez, Alkaline direct alcohol fuel cells, *J. Power Sources* 195 (2010) 3431–3450. https://doi.org/10.1016/j.jpowsour.2009.11.145.

[75] Y.H. Chu, Y.G. Shul, Combinatorial investigation of Pt–Ru–Sn alloys as an anode electrocatalysts for direct alcohol fuel cells, *Int. J. Hydrogen Energy* 35 (2010) 11261–11270. https://doi.org/10.1016/j.ijhydene.2010.07.062.

[76] K. Matsuoka, Y. Iriyama, T. Abe, M. Matsuoka, Z. Ogumi, Alkaline direct alcohol fuel cells using an anion exchange membrane, *J. Power Sources* 150 (2005) 27–31. https://doi.org/10.1016/j.jpowsour.2005.02.020.

[77] L. An, T.S. Zhao, J.B. Xu, A bi-functional cathode structure for alkaline-acid direct ethanol fuel cells, *Int. J. Hydrogen Energy* 36 (2011) 13089–13095. https://doi.org/10.1016/j.ijhydene.2011.07.025.

[78] S.Y. Shen, T.S. Zhao, Q.X. Wu, Product analysis of the ethanol oxidation reaction on palladium-based catalysts in an anion-exchange membrane fuel cell environment, *Int. J. Hydrogen Energy* 37 (2012) 575–582. https://doi.org/10.1016/j.ijhydene.2011.09.077.

[79] N. Wongyao, A. Therdthianwong, S. Therdthianwong, Performance of direct alcohol fuel cells fed with mixed methanol/ethanol solutions, *Energy Convers. Manag.* 52 (2011) 2676–2681. https://doi.org/10.1016/j.enconman.2011.01.005.

[80] Q. Xu, T.S. Zhao, W.W. Yang, R. Chen, A flow field enabling operating direct methanol fuel cells with highly concentrated methanol, *Int. J. Hydrogen Energy* 36 (2011) 830–838. https://doi.org/10.1016/j.ijhydene.2010.09.026.

[81] K.S. Roelofs, T. Hirth, T. Schiestel, Dihydrogenimidazole modified silica-sulfonated poly(ether ether ketone) hybrid materials as electrolyte membranes for direct ethanol fuel cells, *Mater. Sci. Eng.* B 176 (2011) 727–735. https://doi.org/10.1016/j.mseb.2011.02.029.

[82] S. Song, W. Zhou, Z. Zhou, L. Jiang, G. Sun, Q. Xin, V. Leontidis, S. Kontou, P. Tsiakaras, Direct ethanol PEM fuel cells: The case of platinum based anodes, *Int. J. Hydrogen Energy* 30 (2005) 995–1001. https://doi.org/10.1016/j.ijhydene.2004.11.006.

[83] L. An, T.S. Zhao, R. Chen, Q.X. Wu, A novel direct ethanol fuel cell with high power density, *J. Power Sources* 196 (2011) 6219–6222. https://doi.org/10.1016/j.jpowsour.2011.03.040.

[84] L. An, T.S. Zhao, S.Y. Shen, Q.X. Wu, R. Chen, Performance of a direct ethylene glycol fuel cell with an anion-exchange membrane, *Int. J. Hydrogen Energy* 35 (2010) 4329–4335. https://doi.org/10.1016/j.ijhydene.2010.02.009.

[85] L. An, T.S. Zhao, S.Y. Shen, Q.X. Wu, R. Chen, Alkaline direct oxidation fuel cell with non-platinum catalysts capable of converting glucose to electricity at high power output, *J. Power Sources* 196 (2011) 186–190. https://doi.org/10.1016/j.jpowsour.2010.05.069.

[86] Y.S. Li, T.S. Zhao, Z.X. Liang, Performance of alkaline electrolyte-membrane-based direct ethanol fuel cells, *J. Power Sources* 187 (2009) 387–392. https://doi.org/10.1016/j.jpowsour.2008.10.132.

[87] M.Z.F. Kamarudin, S.K. Kamarudin, M.S. Masdar, W.R.W. Daud, Review: Direct ethanol fuel cells, *Int. J. Hydrogen Energy* 38 (2013) 9438–9453. https://doi.org/10.1016/j.ijhydene.2012.07.059.

[88] S.Y. Shen, T.S. Zhao, Q.X. Wu, Product analysis of the ethanol oxidation reaction on palladium-based catalysts in an anion-exchange membrane fuel cell environment, *Int. J. Hydrogen Energy* 37 (2012) 575–582. https://doi.org/10.1016/j.ijhydene.2011.09.077.

[89] N. Wongyao, A. Therdthianwong, S. Therdthianwong, Performance of direct alcohol fuel cells fed with mixed methanol/ethanol solutions, *Energy Convers. Manag.* 52 (2011) 2676–2681. https://doi.org/10.1016/j.enconman.2011.01.005.

[90] Q. Xu, T.S. Zhao, W.W. Yang, R. Chen, A flow field enabling operating direct methanol fuel cells with highly concentrated methanol, *Int. J. Hydrogen Energy* 36 (2011) 830–838. https://doi.org/10.1016/j.ijhydene.2010.09.026.

[91] K.S. Roelofs, T. Hirth, T. Schiestel, Dihydrogenimidazole modified silica-sulfonated poly(ether ether ketone) hybrid materials as electrolyte membranes for direct ethanol fuel cells, *Mater. Sci. Eng. B* 176 (2011) 727–735. https://doi.org/10.1016/j.mseb.2011.02.029.

[92] S. Song, W. Zhou, Z. Zhou, L. Jiang, G. Sun, Q. Xin, V. Leontidis, S. Kontou, P. Tsiakaras, Direct ethanol PEM fuel cells: The case of platinum based anodes, *Int. J. Hydrogen Energy* 30 (2005) 995–1001. https://doi.org/10.1016/j.ijhydene.2004.11.006.

[93] L. An, T.S. Zhao, R. Chen, Q.X. Wu, A novel direct ethanol fuel cell with high power density, *J. Power Sources* 196 (2011) 6219–6222. https://doi.org/10.1016/j.jpowsour.2011.03.040.

[94] L. An, T.S. Zhao, S.Y. Shen, Q.X. Wu, R. Chen, Performance of a direct ethylene glycol fuel cell with an anion-exchange membrane, *Int. J. Hydrogen Energy* 35 (2010) 4329–4335. https://doi.org/10.1016/j.ijhydene.2010.02.009.

[95] L. An, T.S. Zhao, S.Y. Shen, Q.X. Wu, R. Chen, Alkaline direct oxidation fuel cell with non-platinum catalysts capable of converting glucose to electricity at high power output, *J. Power Sources* 196 (2011) 186–190. https://doi.org/10.1016/j.jpowsour.2010.05.069.

[96] Y.S. Li, T.S. Zhao, Z.X. Liang, Performance of alkaline electrolyte-membrane-based direct ethanol fuel cells, *J. Power Sources* 187 (2009) 387–392. https://doi.org/10.1016/j.jpowsour.2008.10.132.

[97] P.W. Gallagher, W.C. Yee, H.S. Baumes, 2015 energy balance for the corn-ethanol industry, *United States Dep. Agric.* (2016) 1–21.

[98] S.M. Haile, Materials for fuel cells, *Mater. Today* 6 (2003) 24–29. https://doi.org/10.1016/S1369-7021(03)00331-6.

[99] H. von Blottnitz, M.A. Curran, A review of assessments conducted on bio-ethanol as a transportation fuel from a net energy, greenhouse gas, and environmental life cycle perspective, *J. Clean. Prod.* 15 (2007) 607–619. https://doi.org/10.1016/j.jclepro.2006.03.002.

[100] L. Luo, E. van der Voet, G. Huppes, An energy analysis of ethanol from cellulosic feedstock–Corn stover, *Renew. Sustain. Energy Rev.* 13 (2009) 2003–2011. https://doi.org/10.1016/j.rser.2009.01.016.
[101] V. Menon, M. Rao, Trends in bioconversion of lignocellulose: Biofuels, platform chemicals & biorefinery concept, *Prog. Energy Combust. Sci.* 38 (2012) 522–550. https://doi.org/10.1016/j.pecs.2012.02.002.
[102] P.S. Nigam, A. Singh, Production of liquid biofuels from renewable resources, *Prog. Energy Combust. Sci.* 37 (2011) 52–68. https://doi.org/10.1016/j.pecs.2010.01.003.
[103] A. Limayem, S.C. Ricke, Lignocellulosic biomass for bioethanol production: Current perspectives, potential issues and future prospects, *Prog. Energy Combust. Sci.* 38 (2012) 449–467. https://doi.org/10.1016/j.pecs.2012.03.002.
[104] V. Guadalupe-Medina, H.W. Wisselink, M.A. Luttik, E. de Hulster, J.-M. Daran, J.T. Pronk, A.J. van Maris, Carbon dioxide fixation by Calvin-Cycle enzymes improves ethanol yield in yeast, *Biotechnol. Biofuels* 6 (2013) 125. https://doi.org/10.1186/1754-6834-6-125.
[105] T.S. Zhao, Z.X. Liang, J.B. Xu, Fuel cells—direct alcohol fuel cells | overview, in *Encyclopedia of Electrochemical Power Sources*, Elsevier, 2009, pp. 362–369. https://doi.org/10.1016/B978-044452745-5.00240-9.
[106] C. Lamy, A. Lima, V. LeRhun, F. Delime, C. Coutanceau, J.-M. Léger, Recent advances in the development of direct alcohol fuel cells (DAFC), *J. Power Sources* 105 (2002) 283–296. https://doi.org/10.1016/S0378-7753(01)00954-5.
[107] S. Eris, Z. Daşdelen, F. Sen, Enhanced electrocatalytic activity and stability of monodisperse Pt nanocomposites for direct methanol fuel cells, *J. Colloid Interface Sci.* 513 (2018) 767–773. https://doi.org/10.1016/j.jcis.2017.11.085.
[108] F. Şen, G. Gökağaç, Different sized platinum nanoparticles supported on carbon: An XPS study on these methanol oxidation catalysts, *J. Phys. Chem. C* 111 (2007) 5715–5720. https://doi.org/10.1021/jp068381b.
[109] Y. Yıldız, E. Erken, H. Pamuk, H. Sert, F. Şen, Monodisperse Pt nanoparticles assembled on reduced graphene oxide: Highly efficient and reusable catalyst for methanol oxidation and dehydrocoupling of dimethylamine-borane (DMAB), *J. Nanosci. Nanotechnol.* 16 (2016) 5951–5958. https://doi.org/10.1166/jnn.2016.11710.
[110] Ö. Karatepe, Y. Yıldız, H. Pamuk, S. Eris, Z. Dasdelen, F. Sen, Enhanced electrocatalytic activity and durability of highly monodisperse Pt@PPy–PANI nanocomposites as a novel catalyst for the electro-oxidation of methanol, *RSC Adv.* 6 (2016) 50851–50857. https://doi.org/10.1039/C6RA06210E.
[111] B. Sen, B. Demirkan, B. Şimşek, A. Savk, F. Sen, Monodisperse palladium nanocatalysts for dehydrocoupling of dimethylamineborane, *Nano-Struct Nano-Objects* 16 (2018) 209–214. https://doi.org/10.1016/j.nanoso.2018.07.008.
[112] B. Sen, E. Kuyuldar, A. Şavk, H. Calimli, S. Duman, F. Sen, Monodisperse ruthenium copper alloy nanoparticles decorated on reduced graphene oxide for dehydrogenation of DMAB, *Int. J. Hydrogen Energy* 44 (2019) 10744–10751. https://doi.org/10.1016/j.ijhydene.2019.02.176.
[113] B. Çelik, Y. Yıldız, H. Sert, E. Erken, Y. Koşkun, F. Şen, Monodispersed palladium–cobalt alloy nanoparticles assembled on poly(N-vinyl-pyrrolidone) (PVP) as a highly effective catalyst for dimethylamine borane (DMAB) dehydrocoupling, *RSC Adv.* 6 (2016) 24097–24102. https://doi.org/10.1039/C6RA00536E.

[114] N. Edwards, S.R. Ellis, J.C. Frost, S.E. Golunski, A.N. van Keulen, N.G. Lindewald, J.G. Reinkingh, On-board hydrogen generation for transport applications: The HotSpot™ methanol processor, *J. Power Sources* 71 (1998) 123–128. https://doi.org/10.1016/S0378-7753(97)02797-3.
[115] M. Baldauf, W. Preidel, Status of the development of a direct methanol fuel cell, *J. Power Sources* 84 (1999) 161–166. https://doi.org/10.1016/S0378-7753(99)00332-8.
[116] A. Demirbas, Direct use of methanol in fuel cells, energy sources, part A recover. *Util. Environ. Eff.* 30 (2008) 529–535. https://doi.org/10.1080/15567030600817159.
[117] H.F. Hunger, The mechanism of oscillatory behavior during the anodic oxidation of formaldehyde, *J. Electrochem. Soc.* 115 (1968) 492. https://doi.org/10.1149/1.2411288.
[118] E. Antolini, Catalysts for direct ethanol fuel cells, *J. Power Sources* 170 (2007) 1–12. https://doi.org/10.1016/j.jpowsour.2007.04.009.
[119] J. Wang, S. Wasmus, R.F. Savinell, Evaluation of ethanol, 1-propanol, and 2-propanol in a direct oxidation polymer-electrolyte fuel cell: A real-time mass spectrometry study, *J. Electrochem. Soc.* 142 (1995) 4218–4224. https://doi.org/10.1149/1.2048487.
[120] L. Jiang, G. Sun, S. Sun, J. Liu, S. Tang, H. Li, B. Zhou, Q. Xin, Structure and chemical composition of supported Pt–Sn electrocatalysts for ethanol oxidation, *Electrochim. Acta* 50 (2005) 5384–5389. https://doi.org/10.1016/j.electacta.2005.03.018.
[121] E. Antolini, E.R. Gonzalez, A simple model to assess the contribution of alloyed and non-alloyed platinum and tin to the ethanol oxidation reaction on Pt–Sn/C catalysts: Application to direct ethanol fuel cell performance, *Electrochim. Acta* 55 (2010) 6485–6490. https://doi.org/10.1016/j.electacta.2010.06.035.
[122] E.E. Altuner, K. Arikan, H. Burhan, S. Ozdemir, F. Şen, Commercial aspects of direct alcohol fuel cells, *Nanomater. Direct Alcohol Fuel Cells Charact. Des. Electrocatal.* (2021) 511–524. https://doi.org/10.1016/B978-0-12-821713-9.00012-3.
[123] W. Qian, D.P. Wilkinson, J. Shen, H. Wang, J. Zhang, Architecture for portable direct liquid fuel cells, *J. Power Sources* 154 (2006) 202–213. https://doi.org/10.1016/j.jpowsour.2005.12.019.
[124] N. Fujiwara, Z. Siroma, S. Yamazaki, T. Ioroi, H. Senoh, K. Yasuda, Direct ethanol fuel cells using an anion exchange membrane, *J. Power Sources* 185 (2008) 621–626. https://doi.org/10.1016/j.jpowsour.2008.09.024.
[125] W. Zhou, B. Zhou, W. Li, Z. Zhou, S. Song, G. Sun, Q. Xin, S. Douvartzides, M. Goula, P. Tsiakaras, Performance comparison of low-temperature direct alcohol fuel cells with different anode catalysts, *J. Power Sources* 126 (2004) 16–22. https://doi.org/10.1016/j.jpowsour.2003.08.009.
[126] S.G. da Silva, M.H.M.T. Assumpção, R.F.B. de Souza, G.S. Buzzo, E.V. Spinacé, A.O. Neto, J.C.M. Silva, Electrochemical and fuel cell evaluation of PtIr/C electrocatalysts for ethanol electrooxidation in Alkaline medium, *Electrocatalysis* 5 (2014) 438–444. https://doi.org/10.1007/s12678-014-0213-2.
[127] J. Goel, S. Basu, Pt-Re-Sn as metal catalysts for electro-oxidation of ethanol in direct ethanol fuel cell, *Energy Procedia* 28 (2012) 66–77. https://doi.org/10.1016/j.egypro.2012.08.041.
[128] J.P. Pereira, D.S. Falcão, V.B. Oliveira, A.M.F.R. Pinto, Performance of a passive direct ethanol fuel cell, *J. Power Sources* 256 (2014) 14–19. https://doi.org/10.1016/j.jpowsour.2013.12.036.

[129] B.C. Ong, S.K. Kamarudin, S. Basri, Direct liquid fuel cells: A review, *Int. J. Hydrogen Energy* 42 (2017) 10142–10157. https://doi.org/10.1016/j.ijhydene.2017.01.117.

[130] T.S. Zhao, Z.X. Liang, J.B. Xu, Fuel cells—direct alcohol fuel cells | Overview, in: *Encyclopedia of Electrochemical Power Sources*, Elsevier, 2009, pp. 362–369. https://doi.org/10.1016/B978-044452745-5.00240-9.

[131] S.P.S. Badwal, S. Giddey, A. Kulkarni, J. Goel, S. Basu, Direct ethanol fuel cells for transport and stationary applications—A comprehensive review, *Appl. Energy* 145 (2015) 80–103. https://doi.org/10.1016/j.apenergy.2015.02.002.

[132] D.M. Fadzillah, S.K. Kamarudin, M.A. Zainoodin, M.S. Masdar, Critical challenges in the system development of direct alcohol fuel cells as portable power supplies: An overview, *Int. J. Hydrogen Energy* 44 (2019) 3031–3054. https://doi.org/10.1016/j.ijhydene.2018.11.089.

[133] G. Li, P.G. Pickup, Analysis of performance losses of direct ethanol fuel cells with the aid of a reference electrode, *J. Power Sources* 161 (2006) 256–263. https://doi.org/10.1016/j.jpowsour.2006.03.071.

[134] S. Rousseau, C. Coutanceau, C. Lamy, J.-M. Léger, Direct ethanol fuel cell (DEFC): Electrical performances and reaction products distribution under operating conditions with different platinum-based anodes, *J. Power Sources* 158 (2006) 18–24. https://doi.org/10.1016/j.jpowsour.2005.08.027.

[135] W.J. Zhou, S.Q. Song, W.Z. Li, Z.H. Zhou, G.Q. Sun, Q. Xin, S. Douvartzides, P. Tsiakaras, Direct ethanol fuel cells based on PtSn anodes: The effect of Sn content on the fuel cell performance, *J. Power Sources* 140 (2005) 50–58. https://doi.org/10.1016/j.jpowsour.2004.08.003.

[136] G.H. Miley, N. Luo, J. Mather, R. Burton, G. Hawkins, L. Gu, E. Byrd, R. Gimlin, P.J. Shrestha, G. Benavides, J. Laystrom, D. Carroll, Direct NaBH4/H2O2 fuel cells, *J. Power Sources* 165 (2007) 509–516. https://doi.org/10.1016/j.jpowsour.2006.10.062.

[137] L. An, T.S. Zhao, R. Chen, Q.X. Wu, A novel direct ethanol fuel cell with high power density, *J. Power Sources* 196 (2011) 6219–6222. https://doi.org/10.1016/j.jpowsour.2011.03.040.

[138] L. Jiang, G. Sun, S. Wang, G. Wang, Q. Xin, Z. Zhou, B. Zhou, Electrode catalysts behavior during direct ethanol fuel cell life-time test, *Electrochem. Commun.* 7 (2005) 663–668. https://doi.org/10.1016/j.elecom.2005.04.021.

[139] F. Takei, N.F. Cooray, K. Yoshida, H. Yoshida, K. Ebisu, S. Suzuki, N. Sawatari, Development of prototype micro fuel cells for mobile electronics, *Fujitsu Sci. Tech. J.* 41 (2005) 191–200.

[140] Toshiba launches Dynario power source for mobile devices, but only in Japan, *Fuel Cells Bull.* 2009 (2009) 6. https://doi.org/10.1016/S1464-2859(09)70394-3.

[141] E.A. Franceschini, H.R. Corti, Applications and durability of direct methanol fuel cells, in *Direct Alcohol Fuel Cells: Materials, Performance, Durability and Applications*, Springer Netherlands, 2013, pp. 321–355.

[142] SFC delivers portable military fuel cell systems, *Fuel Cells Bull.* 2007:7–8. https://doi.org/10.1016/S1464-2859(07)70022-6.

[143] SFC unveils portable Jenny 1200 generator for defence market, *Fuel Cells Bull.* 2013;2013:7–8. ISSN 1464-2859, https://doi.org/10.1016/S1464-2859(13)70320-1.

[144] SFC German order to investigate DMFC use in military vehicles, *Fuel Cells Bull.* 2015;2015:4. https://doi.org/10.1016/S1464-2859(15)70008-8.

[145] U. Schramm, *First Series of Portable JENNY 1200 Fuel Cell Generator Delivered to U.S. Air Force*, Brunnthal, Germany: SFC Energy.
[146] Ballard ships 500th methanol telecom unit, hits record production, *Fuel Cells Bull.* 2013;2013:4. ISSN 1464-2859, https://doi.org/10.1016/S1464-2859(13)70127-5.
[147] M. Müller, N. Kimiaie, A. Glüsen, Direct methanol fuel cell systems for backup power—influence of the standby procedure on the lifetime, *Int J Hydrogen Energy* 2014;39:21739e45.
[148] Ballard in telecom pilot at Idea Cellular in India, *Fuel Cells Bull.* 2013;2013:12 1. ISSN 1464-2859, https://doi.org/10.1016/S1464-2859(13)70397-3.
[149] K. Williamson, Fuel cell auxiliary power proving efficient for Yorkshire emergency response vehicles. *Fuel Cells Bull.* 2011(10) 12–14.
[150] J.H. Hales, C. Kallesøe, T. Lund-Olesen, A.C. Johansson, H.C. Fanøe, Y. Yu, et al. Micro fuel cells power the hearing aids of the future, *Fuel Cells Bull.* 2012;2012:12e6.
[151] S. Malhotra, Onboard battery charging with Oorja's DMFC for material handling vehicles. *Fuel Cells Bull.* 2012;2012:12e5.
[152] J. Andújar, F. Segura, Fuel cells: History and updating. A walk along two centuries. *Renew. Sustain. Energy Rev.* 2009;13:2309e22.
[153] N. Shaari, S.K. Kamarudin, R. Bahru, S.H. Osman, N.A.I. Md Ishak, Progress and challenges: Review for direct liquid fuel cell. *Int. J. Energy Res.* 2021; 45(5) 6644–6688. https://doi.org/10.1002/ER.6353.
[154] A. Islam, T. Islam, H. Mahmud, O. Raihan, M.S. Islam, H.M. Marwani, . . . M.R. Awual, Accelerating the green hydrogen revolution: A comprehensive analysis of technological advancements and policy interventions. *Int. J. Hydrogen Energy* 2024;67 458–486.
[155] G.P. Ling, *Thai Students Beat Own Record, Achieve Highest Mileage at Shell Eco-Marathon Asia*, Malaysia: National Geographic, 2012.
[156] Neah Silent, Falcon partner to integrate fuel cells into UAVs, *Fuel Cells Bull.* 2014;2014:6.
[157] M.S. Alias, S.K. Kamarudin, A.M. Zainoodin, M.S. Masdar, Active direct methanol fuel cell: An overview, *Int. J. Hydrogen Energy* 45 (2020) 19620–19641. https://doi.org/10.1016/j.ijhydene.2020.04.202.
[158] J.I. Martins, M.C. Nunes, R. Koch, L. Martins, M. Bazzaoui, Electrochemical oxidation of borohydride on platinum electrodes: The influence of thiourea in direct fuel cells, *Electrochim. Acta* 52 (2007) 6443–6449. https://doi.org/10.1016/j.electacta.2007.04.066.
[159] V. Gogel, T. Frey, Z. Yongsheng, K. Friedrich, L. Jörissen, J. Garche, Performance and methanol permeation of direct methanol fuel cells: Dependence on operating conditions and on electrode structure, *J. Power Sources* 127 (2004) 172–180. https://doi.org/10.1016/j.jpowsour.2003.09.035.
[160] L.-S. Zhang, X.-Q. Liang, W.-G. Song, Z.-Y. Wu, Identification of the nitrogen species on N-doped graphene layers and Pt/NG composite catalyst for direct methanol fuel cell, *Phys. Chem. Chem. Phys.* 12 (2010) 12055. https://doi.org/10.1039/c0cp00789g.
[161] G. Liu, M. Wang, Y. Wang, F. Ye, T. Wang, Z. Tian, X. Wang, Anode catalyst layer with novel microstructure for a direct methanol fuel cell, *Int. J. Hydrogen Energy* 37 (2012) 8659–8663. https://doi.org/10.1016/j.ijhydene.2012.02.040.

[162] S.-S. Hsieh, H.-C. Wu, B.-S. Her, A novel design for a flow field configuration, of a direct methanol fuel cell, *J. Power Sources* 195 (2010) 3224–3230. https://doi.org/10.1016/j.jpowsour.2009.11.119.
[163] R. Rashidi, I. Dincer, G.F. Naterer, P. Berg, Performance evaluation of direct methanol fuel cells for portable applications, *J. Power Sources* 187 (2009) 509–516. https://doi.org/10.1016/j.jpowsour.2008.11.044.
[164] P.P. Patel, M.K. Datta, P.H. Jampani, D. Hong, J.A. Poston, A. Manivannan, P.N. Kumta, High performance and durable nanostructured TiN supported Pt50–Ru50 anode catalyst for direct methanol fuel cell (DMFC), *J. Power Sources* 293 (2015) 437–446. https://doi.org/10.1016/j.jpowsour.2015.05.051.
[165] A.M. Zainoodin, S.K. Kamarudin, W.R.W. Daud, Electrode in direct methanol fuel cells, *Int. J. Hydrogen Energy* 35 (2010) 4606–4621. https://doi.org/10.1016/j.ijhydene.2010.02.036.
[166] X. Cheng, Characterization of catalysts and membrane in DMFC lifetime testing, *ECS Proc.* 2004–21 (2004) 402–411. https://doi.org/10.1149/200421.0402PV.
[1] L. Jiang, G. Sun, S. Wang, G. Wang, Q. Xin, Z. Zhou, B. Zhou, Electrode catalysts behavior during direct ethanol fuel cell life-time test, *Electrochem. Commun.* 7 (2005) 663–668. https://doi.org/10.1016/j.elecom.2005.04.021.
[167] V. Velisala, G.N. Srinivasulu, B.S. Reddy, K.V.K. Rao, Review on challenges of direct liquid fuel cells for portable application, *World J. Eng.* 12 (2015) 591–606. https://doi.org/10.1260/1708-5284.12.6.591.
[168] G.L. Soloveichik, Liquid fuel cells, *Beilstein J. Nanotechnol.* 5 (2014) 1399–1418. https://doi.org/10.3762/bjnano.5.153.
[169] B.C. Ong, S.K. Kamarudin, S. Basri, Direct liquid fuel cells: A review, *Int. J. Hydrogen Energy* 42 (2017) 10142–10157. https://doi.org/10.1016/j.ijhydene.2017.01.117.
[170] B.C. Ong, S.K. Kamarudin, S. Basri, Direct liquid fuel cells: A review, *Int. J. Hydrogen Energy* 42 (2017) 10142–10157. https://doi.org/10.1016/j.ijhydene.2017.01.117.
[171] S.P.S. Badwal, S. Giddey, A. Kulkarni, J. Goel, S. Basu, Direct ethanol fuel cells for transport and stationary applications—A comprehensive review, *Appl. Energy* 145 (2015) 80–103. https://doi.org/10.1016/j.apenergy.2015.02.002.
[172] R. Chetty, K. Scott, Direct ethanol fuel cells with catalysed metal mesh anodes, *Electrochim. Acta* 52 (2007) 4073–4081. https://doi.org/10.1016/j.electacta.2006.11.043.
[173] E. Antolini, F. Colmati, E.R. Gonzalez, Effect of Ru addition on the structural characteristics and the electrochemical activity for ethanol oxidation of carbon supported Pt–Sn alloy catalysts, *Electrochem. Commun.* 9 (2007) 398–404. https://doi.org/10.1016/j.elecom.2006.10.012.
[174] C. Xu, L. Cheng, P. Shen, Y. Liu, Methanol and ethanol elec trooxidation on Pt and Pd supported on carbon microspheres in alkaline media, *Electrochem. Commun.* 9 (2007) 997–1001. https://doi.org/10.1016/j.elecom.2006.12.003.
[175] D.D. James, P.G. Pickup, Effects of crossover on product yields measured for direct ethanol fuel cells, *Electrochim. Acta* 55 (2010) 3824–3829. https://doi.org/10.1016/j.electacta.2010.02.007.
[176] W. Vielstich, A. Lamm, H.A. Gasteiger, H. Yokokawa, eds., *Handbook of Fuel Cells*, Wiley, 2010. https://doi.org/10.1002/9780470974001.

[177] Z. Liu, X.Y. Ling, X. Su, J.Y. Lee, L.M. Gan, Preparation and characterization of Pt/C and PtRu/C electrocatalysts for direct ethanol fuel cells, *J. Power Sources* 149 (2005) 1–7. https://doi.org/10.1016/j.jpowsour.2005.02.009.
[178] J.J. Linares, S.C. Zignani, T.A. Rocha, E.R. Gonzalez, Ethanol oxidation on a high temperature PBI-based DEFC using Pt/C, PtRu/C and Pt3Sn/C as catalysts, *J. Appl. Electrochem.* 43 (2013) 147–158. https://doi.org/10.1007/s10800-012-0496-z.
[179] J.J. Linares, S.C. Zignani, T.A. Rocha, E.R. Gonzalez, Ethanol oxidation on a high temperature PBI-based DEFC using Pt/C, PtRu/C and Pt3Sn/C as catalysts, *J. Appl. Electrochem.* 43 (2013) 147–158. https://doi.org/10.1007/s10800-012-0496-z.

5

Advances in Membrane and Catalyst in Direct Formic Acid Fuel Cells (DFAFCs)

Abid Ullah, Basharat Hussain, Muhammad Humayun, Muhammad Raheel Khan, and Sanaullah Qamar

5.1 Introduction

Energy functions as the universally accepted currency of technology. The absence of energy would dismantle the very structure of society as we understand it. The impact of a 24-hour power outage in a city starkly highlights our profound reliance on this incredibly valuable resource. Computers halt, elevators freeze, hospitals operate at basic levels, and darkness pervades. As populations grow, often surpassing the average growth rate of 2%, the demand for energy intensifies. Affluent industrialized nations, which only include 25% of people on the planet, consume a startling 75% of the energy used worldwide, demonstrating the correlation between improved lives and increased energy needs [1–3]. Beyond global warming, the problems caused by energy supply and consumption include serious environmental issues such air pollution, acid rain, ozone layer loss, deforestation, and radioactive material emissions. These concerns must be tackled simultaneously if mankind is to ensure a promising energy future with little impact on the environment. Substantial evidence underscores the negative impact we face if environmental degradation continues. The energy sector and the public are increasingly recognizing other environmental considerations. The notion that customers are responsible for some of the pollution and its associated costs is gaining traction, over the past few decades, as the cost of many energy resources has increased in certain jurisdictions to account for environmental factors. By the middle of the twenty-first century, the population of the globe is projected to double, and economic growth is anticipated to continue. Global demand for energy services could multiply by a factor of two by 2050, whereas primary energy needs could rise by a factor of 1.5 to 3 [4,5]. Worries regarding energy-associated sustainability issues, including

 DOI: 10.1201/9781003512592-5

acid rain, depletion of ozone in the stratospheric, and global climate change, are projected to grow along with this course.

In response to these environmental challenges and the surging energy demand, researchers are diligently striving to develop renewable energy sources by converting various fuels into electricity. This conversion necessitates equipment that can seamlessly facilitate the transition from fuel to electricity. Recent developments in equipment, such as fuel cells and batteries, are geared towards establishing the infrastructure required for future energy demands. Among these, fuel cells are gaining prominence due to their high energy densities. Various types of fuel cells are available, including solid oxide fuel cells, polymer electrolyte membrane fuel cells (PEMFCs), alkaline fuel cells, and more, and are being concurrently explored. PEMFCs have garnered significant attention due to their user-friendly nature and relatively low operating temperatures. This growing landscape of energy storage and generation is undergoing a remarkable shift driven by the emergence of fuel cells based on polymer electrolyte membranes (PEMs). This paradigm shift holds the potential to supplant traditional batteries in portable devices. Within this domain, two prominent contenders have long captured our attention: hydrogen-fed PEM fuel cells (PEMFCs) and direct methanol fuel cells (DMFCs) [6–11]. However, despite years of dedicated research, intrinsic limitations persist. Hydrogen-based PEMFCs encounter hurdles rooted in the substantial costs associated with containing small amounts of hydrogen, safety concerns linked to its transportation, and its relatively low energy density when in a gaseous state. In contrast, while DMFCs boast the impressive energy density of liquid methanol (approximately 4900 Wh per litre), they grapple with the challenge of converting it into electricity as efficiently as hydrogen. Moreover, compatibility issues between methanol and Nafion membranes confine these cells to low concentrations (typically between 1 and 2 M). Any deviation from this concentration results in notable fuel crossover, compromising both fuel efficiency and cell performance. Additionally, concerns about the dangers of methanol, especially when it is vaporized, cast doubts on the practical feasibility of commercializing DMFC technology.

The difficulties associated with using hydrogen and methanol have increased interest in a viable replacement referred to as DFAFCs. According to the US Food and Drug Administration, formic acid has gained recognition as a safe food additive due to its extraordinary stability, which allows it to remain liquid at room temperature. In contrast to methanol, formic acid shows less crossover over Nafion membranes. This unique characteristic paves the path for the use of thinner membranes and fuel solutions in DFAFCs that are extremely concentrated. Formic acid is a good candidate for an electrolyte since it retains its liquid state at ambient temperature. characterized by a well-controlled fuel permeation and impressive power output at standard temperatures, the formic acid-oxygen direct fuel cells have an exceptionally high theoretical open circuit potential of 1.45 V. Additionally,

DFAFCs outperform both hydrogen and direct methanol fuel cells in terms of electric potential difference (EPD), as determined using Gibbs free energy.

DFAFC technology has demonstrated outstanding electrocatalytic capacities in a relatively short period of research, frequently outperforming DMFCs and, in some cases, even matching the performance of hydrogen-based PEMFCs [12]. While formic acid has a volumetric energy density of 2104 Wh per litre, this constraint can be circumvented by employing highly concentrated formic acid solutions. The benefits of DFAFCs can outweigh those of its major equivalent, the DMFC, particularly in smaller power systems. Notably, direct formic acid fuel cells using polymer electrolyte membranes demonstrate attractive qualities for micropower production [13–17]. Protons can move more easily inside the anode chamber of PEM fuel cells thanks to formic acid's natural conductivity and its function as a rather powerful electrolyte. This unique characteristic, which is caused by an anodic repulsion mechanism between Nafion and the partially dissociated form of formic acid (formate anions), results in limited fuel crossover from the anode to the cathode. Although formic acid's power density is lower than methanol's at 1740 Wh per kilogram, it still ranks as a strong competitor in terms of fuel power density when used in a highly concentrated 20 M (70% by weight) formic acid solution.

This chapter is dedicated to exploring the recent strides in DFAFC research, with a primary focus on the advancements in catalysts for the conversion of formic acid into electricity and the investigation of alternative membrane materials. Additionally, we delve into the fundamental chemistry of DFAFCs, including the intriguing phenomenon of formic acid crossover through Nafion membranes. Furthermore, we examine the evolution in the design of DFAFC configurations. Through these comprehensive discussions, our objective is to offer a thorough insight into the significant progress achieved in the field of DFAFCs and illuminate their promising prospects.

5.2 Mechanism of Formic Acid Electrocatalytic Oxidation

In alignment with the functioning of various fuel cells employing polymer electrolyte membranes, the direct formic acid fuel cell operates through an air cathode. Within the cathode, the process of oxygen reduction transpires, often involving a platinum-based catalyst that aids in facilitating a four-electron reaction. Meanwhile, at the anode, the direct oxidation of formic acid results in the liberation of two electrons per individual molecule. A comprehensive understanding of the reactions transpiring at the cathode, anode, and their collective impact in a direct formic acid fuel cell can be elucidated as follows:

(1) Anode:

$HCOOH \rightarrow CO_2 + 2H^+ + 2e^-$

with an electrode potential (E_0) of approximately −0.25 V (vs. standard hydrogen electrode, SHE)

(2) Cathode:

$1/2O_2 + 2H^+ + 2e- \rightarrow H_2O$

with an E_0 of 1.23 V (vs. SHE)

(3) Overall:

$HCOOH + 1/2O_2 \rightarrow CO_2 + H_2O$

generating an open circuit voltage (OCV) of about 1.48 V.

Due to its much greater open circuit voltage (OCV) and electromotive force (EMF), the direct formic acid fuel cell stands out, which is roughly 1.48 V [18]. Surprisingly, this EMF is greater than that of direct methanol and hydrogen fuel cells. Formic acid's theoretical energy density is determined using the formula F = 96,485 C moll, which takes into account the compound's molecular weight (MW), OCV, and Faraday constant. Formic acid's inherent energy density according to this calculation is 1725 Wh kg^{-1}. Formic acid has a density of 1.22 kg L^{-1} when it is liquid, which corresponds to a hypothetical energy density of 2104 Wh L^{-1}.

While research on formic acid fuel cells has been ongoing for roughly 12 years (the first paper on formic acid as a fuel for PEM fuel cells published in 1996), even further back in history is the study of the process underpinning formic acid oxidation. This mechanism has been under scrutiny for a substantial duration [19,20]. Among the diverse proposals, the most widely embraced explanation is referred to as the "parallel or dual pathway mechanism." This mechanism introduces two distinct pathways for the oxidation process:

In the initial pathway (Pathway 1), direct oxidation occurs via a dehydrogenation reaction, omitting the formation of carbon monoxide (CO) as an intermediate step:

(4) Route 1: $HCOOH \rightarrow CO_2 + 2H^+ + 2e^-$

In the second route, an intermediate is generated involving through a dehydration process, adsorbed carbon monoxide (CO):

(5) Route 2: $HCOOH \rightarrow -CO_{ads} + H_2O \rightarrow CO_2 + 2H^+ + 2e^-$

In the alternate route, formic acid binds to the catalyst surface, resulting in the creation of an intermediate adsorbed CO species. Subsequently, this intermediate species is oxidized, leading to the formation of gaseous CO_2. In the context of direct formic acid fuel cells, the preferred reaction route is the dehydrogenation pathway, which not only boosts overall cell efficiency but also safeguards the catalyst from potential poisoning. The selection of anode catalyst plays a pivotal role in steering the formic acid oxidation towards the first reaction pathway. A visual representation of a typical DFAFC encompasses the essential components illustrated in Figure 5.1:

Anode: This is where formic acid is introduced to the fuel cell. Formic acid undergoes oxidation at the anode, releasing electrons and protons. Cathode: The cathode is where oxygen from the air is supplied. Oxygen reduction occurs at the cathode, combining with protons and electrons from the anode to form water. Electrolyte: The polymer electrolyte membrane (PEM) is positioned between the anode and the cathode. It allows the passage of protons (H^+ ions) while preventing the mixing of the fuel and oxygen gases. Catalysts: Catalysts are used to facilitate the electrochemical reactions at the anode and cathode. Typically, platinum-based catalysts are used for both the anode and cathode reactions. Electron flow: Electrons released during the oxidation of formic acid at the anode travel through an external circuit to the cathode, creating an electrical current. Proton flow: The PEM transports formic acid's produced protons to the cathode, where they mix with oxygen and electrons to create water. Output: The electrical current generated by the DFAFC can be used to power electronic devices or perform other useful work. Overall reaction: The direct oxidation of formic acid and the reduction of oxygen, which result in the production of carbon dioxide and water, respectively, constitute the overall reaction of the DFAFC.

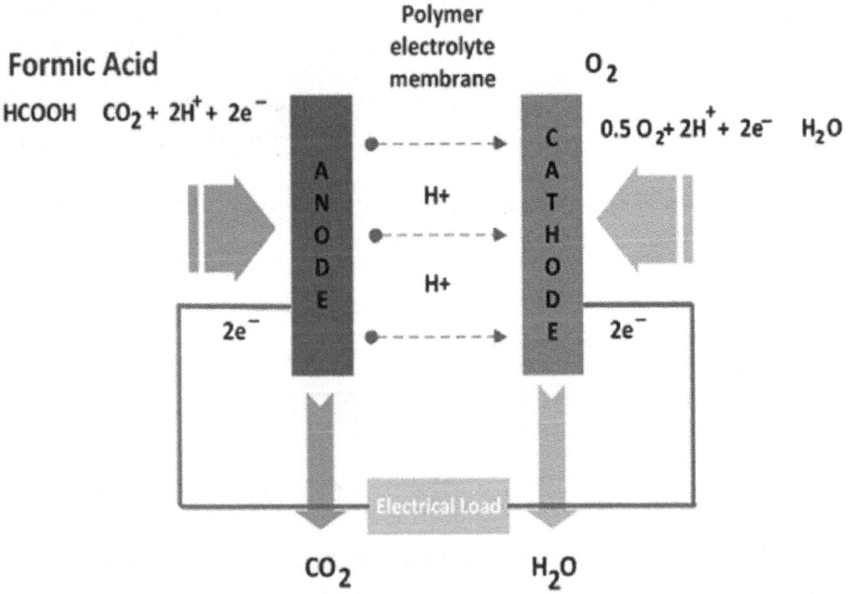

FIGURE 5.1
Schematic representation of DFAFC working procedure.

5.2.1 Formic Acid Electro-Oxidation Catalysts

5.2.1.1 Platinum-Based Catalysts

During the initial phases of DFAFC advancement, catalysts based on platinum (Pt) were prominently utilized within the anode layer. In an early milestone report by Weber et al. [21], using a combination of Pt-black and Pt/Ru catalysts, formic acid was found to have a higher electrolytic activity than methanol. Furthermore, the Pt/Ru catalyst exhibited heightened activity in formic acid oxidation in comparison to Pt-black. The summarized outcomes of these investigations are presented in Figure 5.2. Despite the significant implications of these findings, surprisingly, there were no subsequent reports on DFAFCs until the year 2002, despite ongoing research into the fundamentals of formic acid electro-oxidation [22].

In 2002, a notable research endeavor resurfaced the assertion that formic acid stands as a commendable fuel candidate for fuel cells and could potentially serve as a compelling alternative for compact, portable applications. In order to accelerate the electrically oxidizing process of formic

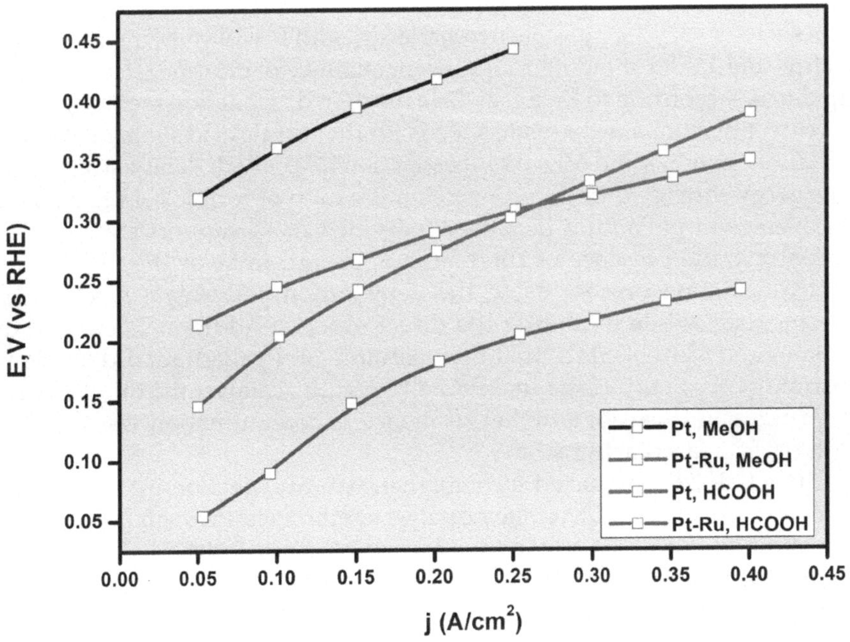

FIGURE 5.2
Pt-black and Pt/Ru anodes, methanol and formic acid oxidation polarization curves were observed. Cathode was Pt-black, 4 mg cm^{-2}, feed rate was 0.26 ml min^{-1}, the electrolyte was a PBI membrane doped with H_3PO_4, the temperature was 170°C, and the air supply rate was 10 ml min^{-1}. The anode catalyst loading was 4 mg cm^{-2}. The ratio of water to fuel was 2. The feed rate was 0.26 ml min^{-1} [21].

acid, a specially developed platinum (Pt)-based catalyst was used in this study's anode layer [12]. The DFAFC, the outcome of this creative endeavor, was first demonstrated with currents up to 134 mA cm^{-2} and power outputs up to 48.8 mW cm^{-2} [12]. Notably, research and interest in bimetallic Pt-M catalysts for DFAFCs have increased [23]. (M designating a different metal from platinum as well as another.) In addition to maximizing the use of platinum, using a second metal improves the performance of the catalyst when oxidizing formic acid. Numerous combinations, such as Pt/Ru, Pt/Pd, Pt/Au, and Pt/Pb, have been investigated as anode catalysts in DFAFCs after thorough analyses of the effects of Ru, Pd, Au, and Pb on catalytic performance. Waszczuk et al. produced Pt-based catalysts in an important study by enabling Pd and/or Ru to spontaneously deposit onto the platinum nanoparticles [24]. The activity was most obviously boosted when Pd was used to decorate the Pt surface. Then Rice et al. thoroughly investigated these Ru- and Pd-adorned Pt catalysts. Figure 5.3 and Table 5.1 compare the DFAFC's performance while using catalysts made of Pt-black, Pt/Ru, and Pt/Pd. Ru and Pd were both shown to considerably boost the catalytic function using Pt/Pd performing better at higher cell voltages. Pt/Ru demonstrated the highest peak power density at 70 mW cm^{-2}, although this was observed at a shockingly low cell potential of 0.26V. Different catalysts showed varying rates of current decay, with Pt/Ru showing the fastest decline and Pt/Pd showing a significant stability of current after an initial rapid loss. According to Rice et al. [23], the dehydrogenation method, which is a direct reaction mechanism, speeds up the formic acid's rate of electron oxidation. Thomas and Masel's investigation [25], which demonstrated that the energy threshold for the conversion the ratio of formic acid to CO_2 fell as coverage of palladium increased, provided more support for this idea. In contrast, the presence of ruthenium appeared to favor the formic acid electro-oxidation process using the dehydration pathway, a reactive CO intermediate, while inhibiting the direct process. Additional research by Thomas and Masel [25] showed that platinum and palladium did not demonstrate synergistic effects in a Pt/Pd bimetallic catalyst for the oxidation of formic acid. Instead, a higher resilience to contamination was credited with the increased activity.

Choi et al. [26] conducted a comprehensive investigation into the viability of an unsupported PtAu alloy catalyst synthesized through borohydride reduction, intended for use as an anode catalyst in the electro-oxidation of formic acid within a DFAFC. In contrast to the more widely used PtRu catalyst, their investigation pitted the PtAu catalyst against it. As seen in Figure 5.4, the PtAu catalyst had greater catalytic activity. It is impressive that the PtAu catalyst, which improved cell performance, maintained effectiveness over a long period of cell operation. According to the authors' hypotheses, the PtAu catalyst oxidizes formic acid primarily through the direct dehydrogenation pathway, which reduces the amount of adsorbed CO that forms [26].

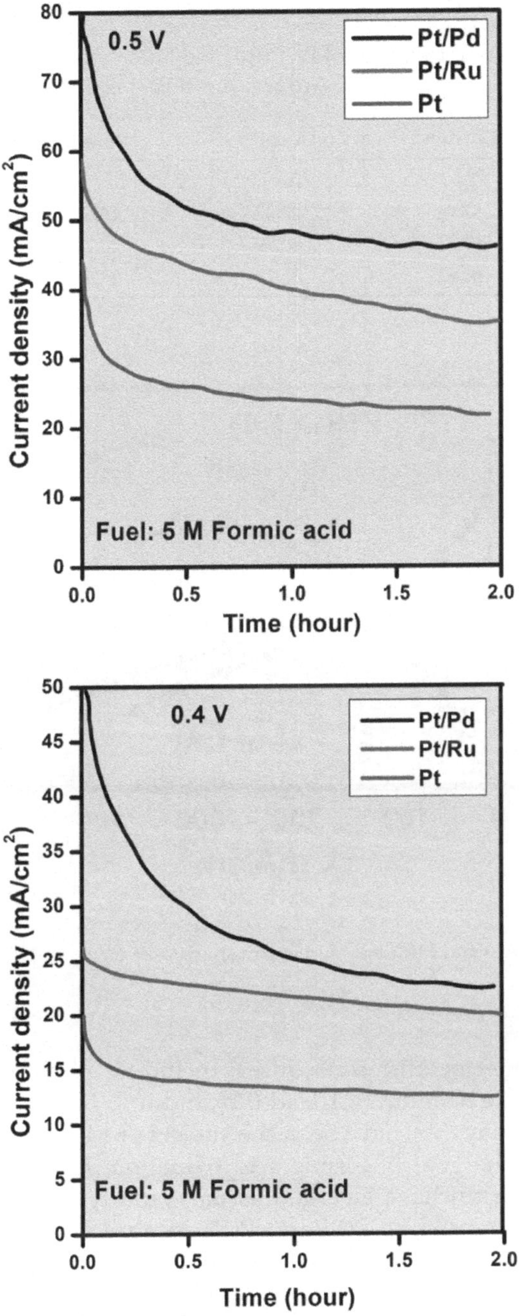

FIGURE 5.3
Constant voltage measurements on a DFAFC at 0.5 and 0.4 V. Fuel: 0.2 ml of 5 M formic acid per minute, platinum black, platinum/ruthenium, and platinum/pd catalysts, and 30°C for the cell [23].

TABLE 5.1

The Power and Current Densities of Pt, PtRu, and PtPd Anode Catalyst-Equipped DFAFCs Operating at 0.4 and 0.5 V Are Measured Respectively [23]

Empty Cell	Current Density (mA cm^{-2})		Power Density (mW cm^{-2})	
Empty Cell	*0.4 V*	*0.5 V*	*0.4 V*	*0.5 V*
Pt	37.44	22.02	15.69	10.30
PtPd	67.32	46.39	28.24	21.71
PtRu	60.61	35.14	25.40	16.44

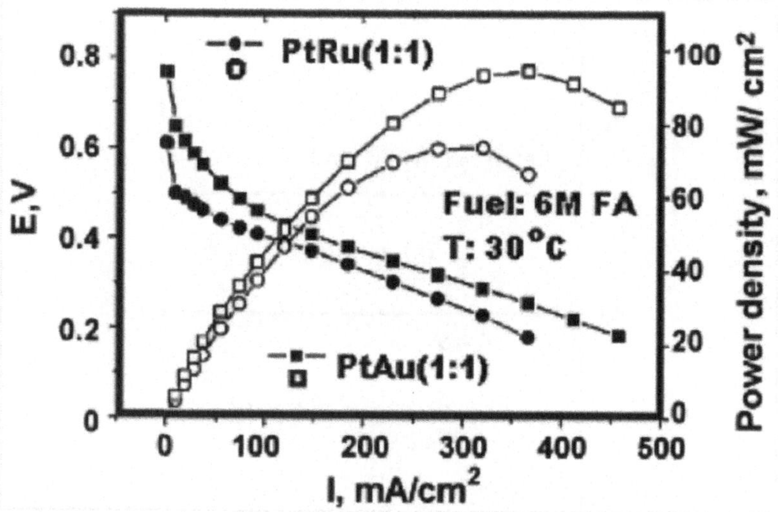

FIGURE 5.4
Performances of PtAu- and PtRu-based MEAs in single cells using 30 °C formic acid and 6 M oxygen [26].

Numerous intermetallic procedures, including platinum (Pt), coupled with elements like bismuth (Bi), lead (Pb), indium (In), tin (Sn), manganese (Mn), and antimony (Sb), have been the subject of substantial investigation with regard to the goal of formic acid oxidation. From this group, PtBi, $PtBi_2$, PtPb, and PtIn have emerged as particularly appealing candidates [27]. Interest in improving PtBi and PtPb catalyst activity has prompted research into surface treatment techniques. Additionally, various methods for producing PtPb nanoparticles have been created and contrasted for effectiveness [28]. A study using an induction-melted PtBi alloy has shed light on the critical roles of Bi dissolution, the deposition of Bi atoms under potential (upd) on the surface, and the creation of Bi oxides. It has been intriguingly demonstrated that surface modification of PtRu with permanently installed Bi atoms can boost formic acid oxidationactivity [29]. By preventing CO from attaching to Pt sites, this is thought to improve catalytic activity [30]. The significant improvement in the formic acid oxidation activity attained with bimetallic PtPb alloys created during arc melting is an intriguing discovery. When compared to pure Pt, this catalyst demonstrates substantially stronger and longer-lasting activity (for over 1,000 seconds) [31]. The investigation includes PtPb nanoparticle catalysts, which not only exhibit increased formic acid oxidation activity but also exhibit improved performance stability over long periods of time (over nine hours) in contrast to Pt, PtRu, and Pd catalysts [30]. When metals like bismuth (Bi) and arsenic (As) are added, formic acid oxidation is said to increase. The fascinating "third-body effect" has been proposed as the underlying process. This result arises from the fact that the availability of CO adsorption sites is reduced when a second element is incorporated into the structure of platinum. Geometrical blockage develops in comparison to a clean Pt surface, making the surface less susceptible to contamination by adsorbed CO [32]. Nevertheless, an electrical interaction between lead (Pb) and platinum (Pt) has been suggested as the cause of the higher activity of PtPb catalysts in the oxidation of formic acid [33]. Potential benefits of electronic processes that act in conjunction with the "third-body effect" may be seen for bi-modified Pt catalysts [34]. The fundamental mechanism governing the oxidation of formic acid has been thoroughly examined using Pt single crystals modified with up to a monolayer of a secondary metal. Arenz et al. conducted a thorough investigation of the Pt (111)-Pd system and single crystals of PtPd alloy, which was published in [35]. At 0.4 V, Pd atoms on the surface were up to three times more active than Pt atoms. It was a fascinating discovery that the CO adsorbed on PT sites but not on PD sites. However, Macia et al. investigated the impact of Bi adlayers on the oxidation of formic acid at both stationary and spinning Pt (111) electrodes. The activity increased as the bi coverage reached 0.1–0.25 of a monolayer, depending on the formic acid concentration. It is significant to notice that current-time transients did not reveal any poisoning symptoms in modified electrodes. According to the study, the amount of adsorbed CO does not significantly increase when the process goes through an intermediate adsorbed form. Particularly remarkable formic acid oxidation activity has been shown in a Pt_4Mo alloy [35]. The risk of CO toxicity was observed to decrease when a hydrous Mo oxide formed on the surface, and it may also hasten the primary route of formic acid oxidation.

A hypothetical perspective, Demirci [36] has suggested the bimetallic PtAg system as a potentially suitable catalyst for DFAFCs. However, in order to offer empirical evidence for this idea, experimental research is required. A fascinating development in a Pt-FeTSPc co-catalyst (tetrasulfophthalocyanine) for formic acid electro-oxidation was made recently [36]. The impact

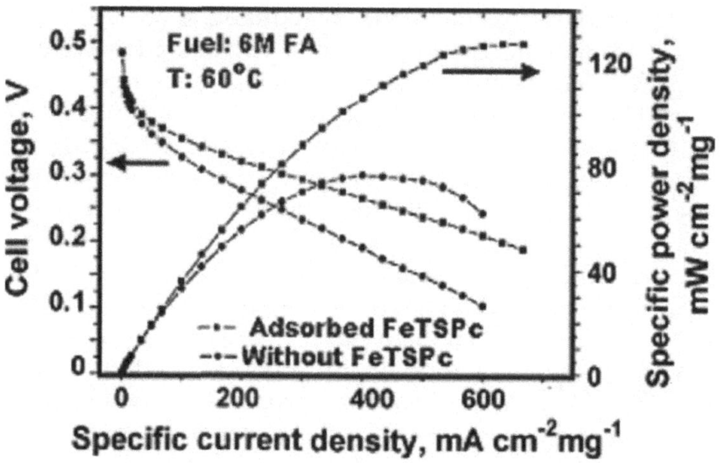

FIGURE 5.5
Pre- and post-FeTSPc anode adsorption polarization patterns for a DFAFC. "Current density/Pt loading" refers to a specific current density. The cell was run with 6 M formic acid at 60°C and air pressure. Cathode: 100 sccm of dry oxygen with a Pt loading of 1.5 mg cm^{-2}; anode: fuel flow rate of 1.0 ml per minute, Pt loading of 0.5 mg cm^{-2} [37].

of utilizing the catalyst made of Pt and FeTSPc on cell performance is succinctly depicted in Figure 5.5. According to the researchers, the modification's improved activity is caused by a mix of intrinsic kinetic enhancement brought on by FeTSPc's donation of electrons and steric hindrance, which prevents the production of adsorbed CO [37].

5.2.2 Pd-Based Catalysts Without Support

There has been a lot of research done on formic acid electro-oxidation on Pd catalysts. It has been stated that DFAFCs use 100% Pd as the anode catalyst. At surrounding (22°C) and higher (30–50°C) conditions, the Pd-black yielded remarkably large power densities when compared to Pt-based catalysts. A typical comparison for Pd-black, PdAu, and PtRu can be found in in Figure 5.6. According to Ha et al.'s research [38], DFAFC performance dropped as the Pd-black catalyst's activity reduced over the course of the cell's operation. By using a strong anodic potential, the lost activity may, however, be largely restored. The same research also looked into the impact of formic acid content and operation temperature on the performance of an H2/air fuel cell, a DMFC (direct methanol fuel cell), a DFAFC with a PtRu anode, and a DFAFC with a Pd anode. Table 5.2 gives a summary of the conclusions. The power density of the Pd anode-catalyzed DFAFC was significantly higher than that of the PtRu catalyst-based DFAFC and DMFC, and it was virtually as effective as the hydrogen-PEM fuel cell. Nevertheless, it was once more shown that restoring the cells initial performance could be

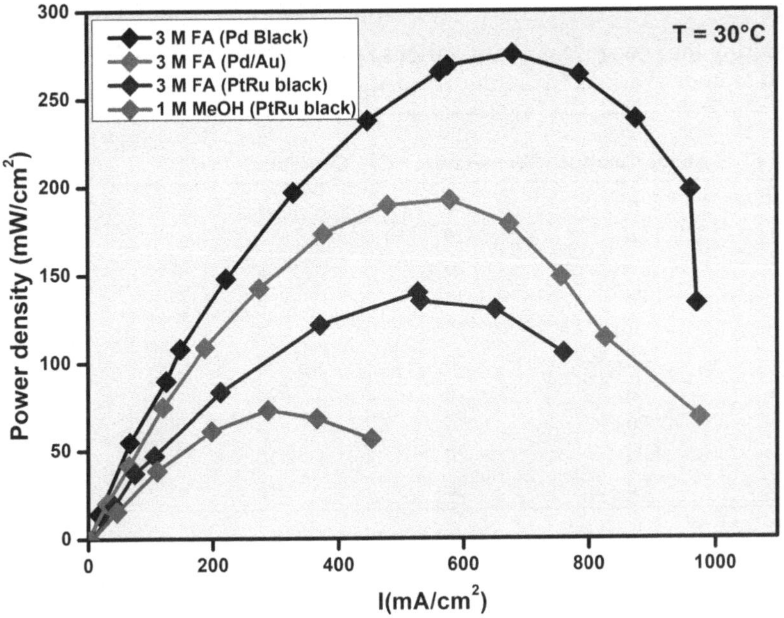

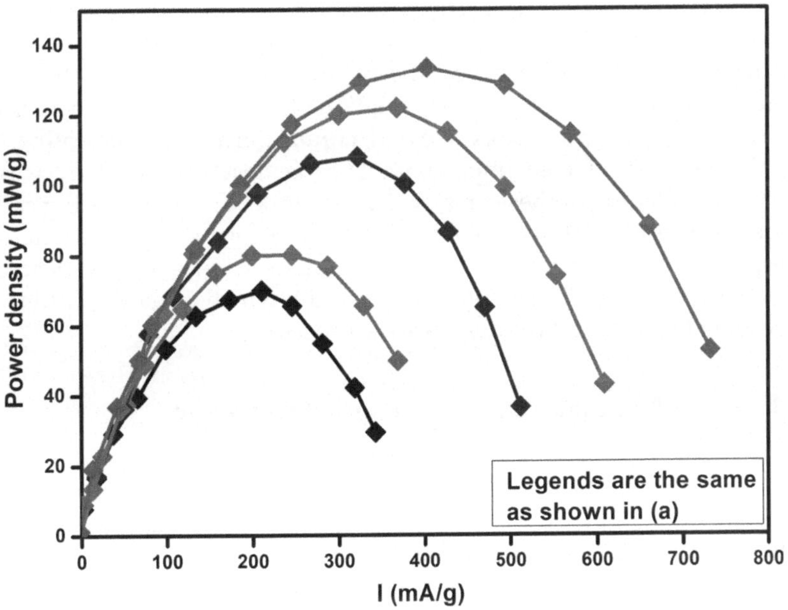

FIGURE 5.6
Power density graphs for fuel cells fed with hydrogen gas, 3 M formic acid, and 1 M methanol. Methanol flow rates were 1 ml/min, formic acid flow rates were 3.5 ml/min, and hydrogen flow rates were 200 sccm, respectively. A total volume of 390 sccm without moisture air was delivered to the cathode. The actual temperature inside the cell is approximately 30°C. For the cathode, a Pt-black catalyst was used, whereas the anodic catalysts utilized in the cell were specified [38].

TABLE 5.2

Comparing the Power Densities of Various Fuel Cell Types and the DFAFCs' Power Densities under Various Operating Scenarios [39]

Fuel Cell	Anode Catalyst	Temperature (°C)	Fuel Concentration (M)	Peak Power Density (mW cm^{-2})
MeOH-air	PtRu	20	1	50
H_2-air	Pd	20	–	320
FA-air	Pd	20	3	253
FA-air	Pd	40	3	350
FA-air	PtRu	20	3	84
FA-air	Pd	30	3	300
FA-air	Pd	20	20	103
FA-air	Pd	50	3	375
FA-air	Pd	20	10	255
FA-air	Pd	20	15	230

accomplished by briefly giving the anode an electrical potential of 1.2 V. This was of particular concern when the DFAFCs were run with elevated formic acid concentrations [39].

Using impedance spectroscopy, recent study has investigated the deactivation of Pd-black in the setting of oxidation of formic acid within a DFAFC [40]. The anode Nyquist plots showed non-uniform arcs, indicating that several processes affected impedance. During deactivation, there was an observable widening of these arcs, which was linked to an escalating charge transfer resistance. Additionally, an increase in formic acid concentration corresponded to an increase in charge transfer resistance. It was speculated that a different organic compound from CO, which usually poisons the anode, may have caused the deactivation process.

5.2.3 Pt- and Pd-Based Catalysts Supported on Carbon

Due to their advantageous characteristics' excellent chemical stability, electrical conductivity, and economic viability, high surface area carbon compounds are commonly used as substrates for precious metal fuel cell catalysts. By using less noble metal, this application hopes to cut the cost of the entire system. Formic acid oxidation kinetics on Pt/C catalysts were remarkably close to those on pure platinum, according to Lovic et al. [41]. Direct dehydrogenation of HCOOH played a major role in the reaction's dual-pathway mechanism, which also resulted in the production of poisonous species. Carbon-supported palladium catalysts have grown significantly in DFAFC research during the past few years. They display noticeable activity, have the ability to use palladium metal more effectively, and enable lower metal loadings [38]. Chemical solution-phase impregnation processes are frequently used to produce Pd/C or Pd-M/C (where M is a different metal from Pd) catalysts. Ha et al. successfully produced palladium loadings of 20% and 40% as uniformly dispersed Pd particles on Vulcan XC-72 carbon support [38]. DFAFCs using these two Pd/C catalysts were able to generate maximum power densities of 145 mW cm^2 (with 20% Pd/C) and 172 mW cm^2 (with 40% Pd/C), as illustrated in Figure 5.7a [42]. The 20% Pd/C catalyst had a lower total power density in a DFAFC when compared to pure palladium, but the amount of electricity it produced per gram of precious metal was significantly higher than that of Pd-black, as shown in Figure 5.7b. Additionally, during cell operation, Pd/C catalysts showed reduced deactivation, particularly at higher formic acid concentrations [38]. Figure 5.7 illustrates how adding gold to Pd/C boosted the catalyst's activity even further, in accordance with Larsen et al.'s findings [42].

Liu et al. [43] employed a microwave-assisted polyol process to create a Pd/C catalyst, which, it is claimed, produced a DFAFC with a high-power density. Figure 5.8 compares the performance of single DFAFCs, employing the microwave-assisted method to produce Pt/C and Pd/C catalysts [43].

Zhang et al. provided a thorough explanation of the manufacture of a Pd/C catalyst employing an aqueous solution phase and the additions of NH_4F and H_3BO_3 [44]. It has been shown that NH4F forms a combination with $PdCl_2$ to produce finely distributed Pd particles with smaller average diameters (approximately 3.2 nm) and considerably poorer crystallinity. As a result, the NH_4F-produced Pd/C catalyst showed improved catalytic performance and durability. Figure 5.9a provides an exemplary representation of these findings. Zhang et al. [44] also published Pd-P/C, another catalyst that was produced utilizing solution phase synthesis with NaH_2PO_2 serving as a co-precursor. As shown in Figure 5.9b, the palladium catalyst's activity and stability were improved after the addition of the non-metal element phosphorus.

Li and Hsing [46] also employed the stabilizer 3-(N,N-dimethyldodecylammonio) propanesulfonate (SB12) as a surfactant to create a PdPt catalyst supported on carbon. The PtPd/C catalysts stabilized with SB12 demonstrated distinguishable advantages over a commercial $Pt_{0.5}Pd_{0.5}$/C catalyst (E-TEK), including superior PtPd dispersion and improved catalytic performance [46]. In a work on PtPb catalysts, Uhm et al. [47] modified platinum-supported electrodes with lead deposited at an underpotential (upd). Compared to pure platinum or PtRu catalysts, the electrocatalytic activity of the PtPbupd catalysts was higher. A Pt/Pbupd/Pt multilayer anode configuration was also used, which showed improved cell performance and stability throughout a five-hour test period [48]. Additionally, it has been found that carbon-supported, Pt-modified gold nanoparticles are efficient at oxidizing formic acid [48].

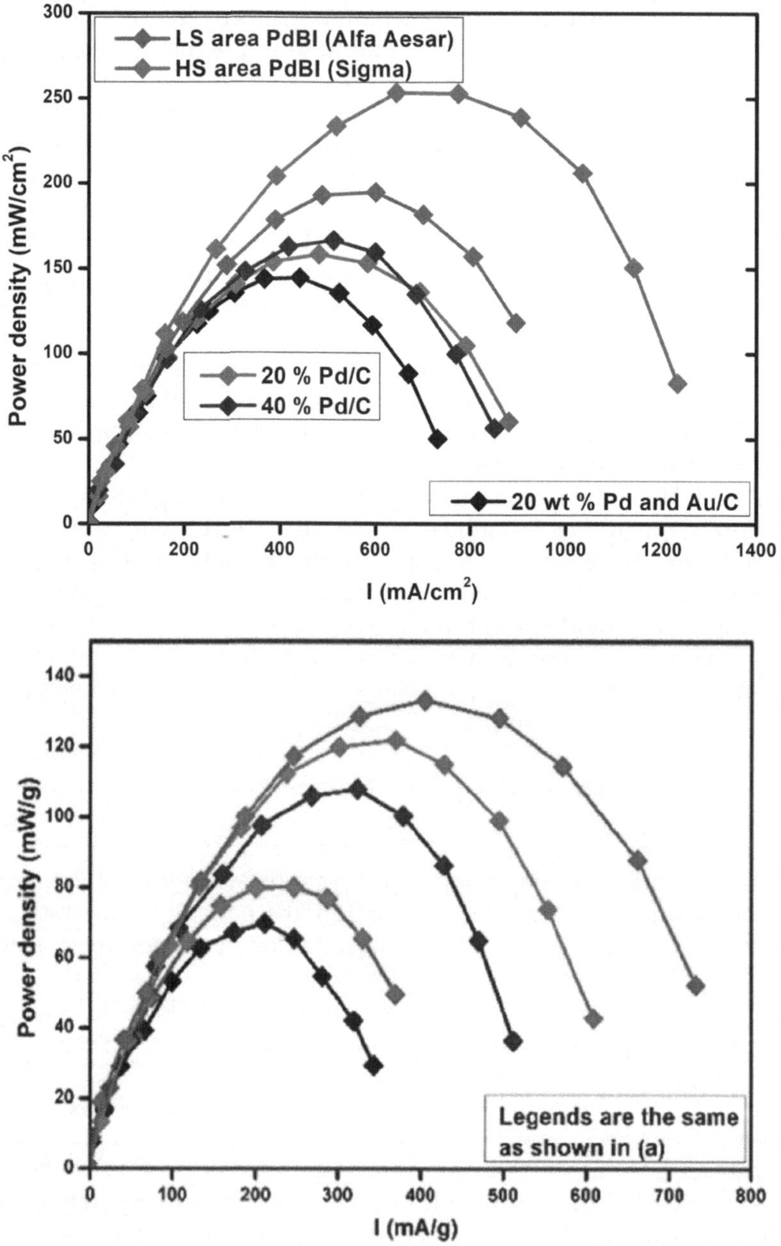

FIGURE 5.7
Power density graphs are shown (a) using several Pd anode catalysts. A direct formic acid fuel cell running in dry air at 30°C with 5 M formic acid was measured (a) per membrane surface area and (b) per total catalyst mass [42].

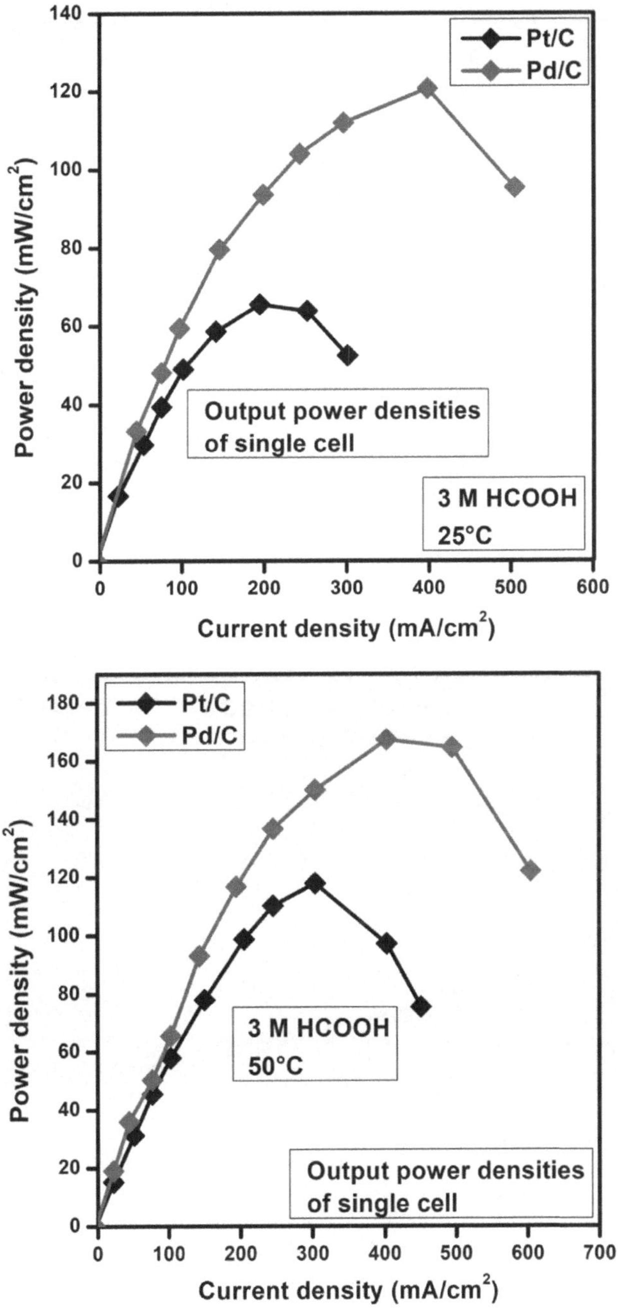

FIGURE 5.8
A single cell performing at (a) 25°C and (b) 50°C generates power densities. Anode: 8 mg cm^{-2} of either platinum or palladium, 3 M HCOOH, and 2 ml/min^{-1}. Cathode: 4 mg cm^{-2} of Pt/C (E-TEK) and 500 cm^3 min^{-1} of O_2 [43].

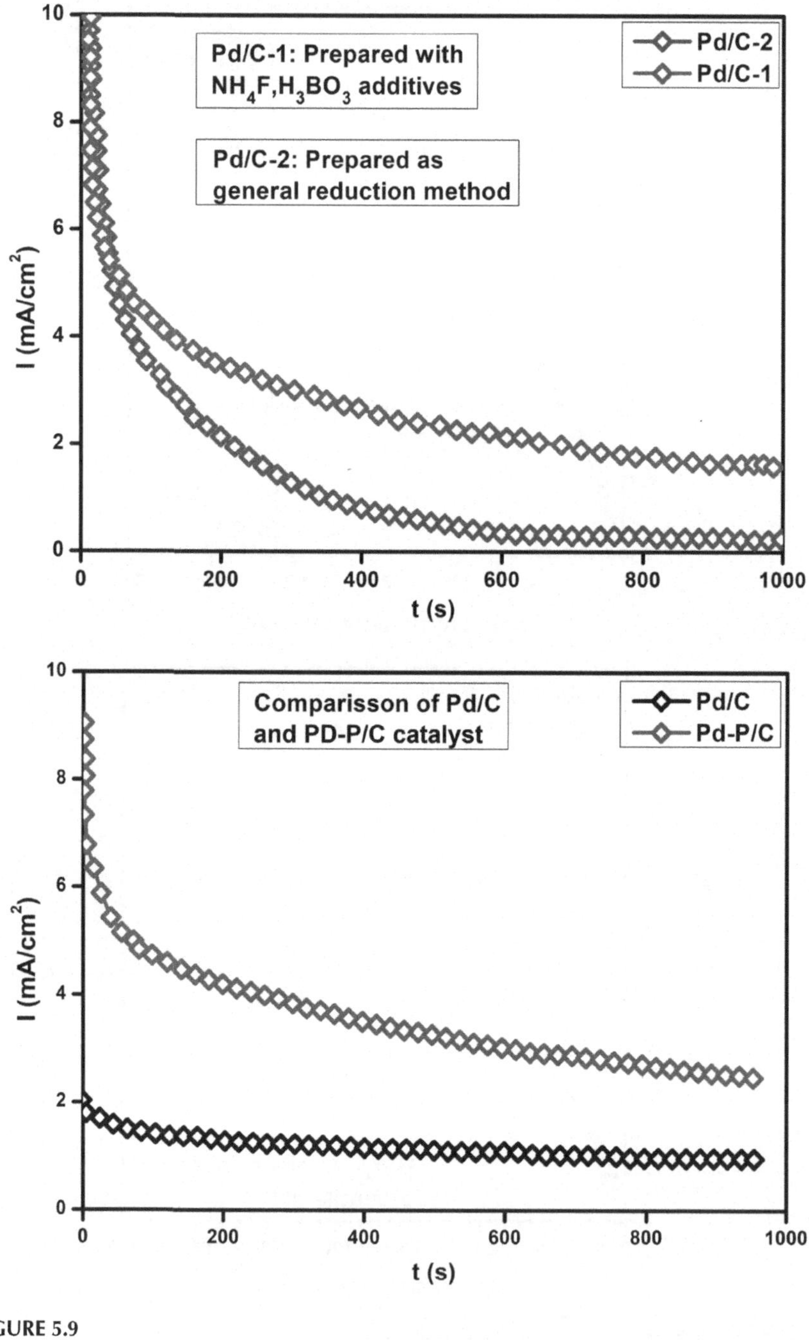

FIGURE 5.9
Chronoamperometric data showing how different Pd/C catalyst electrodes oxidized 0.5 M HCOOH in 0.5 M H_2SO_4 solution. (a) Evaluation of Pd/C catalysts made with and without NH_4F and H_3BO_3 additives. Fixed potential: 0.10 V [45]; (b) catalytic electrodes built of Pd/C and Pd-P/C are contrasted. 0.3 V potential fixed [44].

5.2.4 Additional Methods of Supporting Pt-Based and Pd-Based Catalysts

Recent studies have looked at other substrates besides carbon for making platinum- or palladium-based catalysts. Yi et al. introduced a pioneering method [49] for the hydrothermal synthesis of a PtIr/Ti nanoporous bimetallic catalyst on titanium support. Formaldehyde served as the reduction agent in this process, with H_2PtCl_6 and $IrCl_3$ serving as the precursors. In contrast to pure platinum, the PtIr catalysts supported on titanium exhibited significantly enhanced catalytic activity for formic acid oxidation. The authors suggested that the electro-oxidation of formic acid on PtIr catalysts followed a dehydration pathway involving an intermediate "CO." In DFAFC environments, it's common for highly active palladium-based catalysts to undergo passivation, leading to a gradual decrease in DFAFC performance. To develop highly durable palladium catalysts, Larsen et al. [50] examined the stability of palladium when deposited as a sub-monolayer on various metal foil supports, such as V, Mo, W, and Au. Among these catalysts supported by metal foil, Pd-V, illustrated in Figure 5.10, exhibited the highest stability. The findings from these Pd/metal foil catalysts, as outlined by the authors, offer valuable insights into the creation of palladium-based catalysts with extremely high stability through the incorporation of a secondary metal phase, such as vanadium.

5.3 Permeation of Formic Acid Across Nafion Membranes

The fuel supplied to the negative electrode (anode) in a PEM-based fuel cell has the ability to pass through the membrane and reach the positive electrode (cathode). Because of this, less fuel is used, creates an unfavorable mixed potential, competes with the cathode catalyst, and may even pollute it [51,52]. As a result, the oxygen reduction reaction is less effective. If liquid fuels are used, it could potentially result in the flooding of the cathode catalyst layer [52]. When used in PEM fuel cells, formic acid is frequently praised for its poor penetration across Nafion membranes. Formic acid crossover, on the other hand, must be kept in mind because it still poses a substantial problem and lowers the efficiency of DFAFCs. Jeong et al. [53] investigated the formic acid crossover behavior within an actual operational DFAFC environment under various conditions. According to study by Wang et al. [54] conducted under non-fuel cell conditions, the formic acid crossover behavior was shown to be consistent with their findings. Figure 5.11 shows that as formic acid is used at the anode, formic acid crossover rises with formic acid concentration while falling with current density. When considered in relation to the crossover current (i.e., the current needed to oxidize the fuel

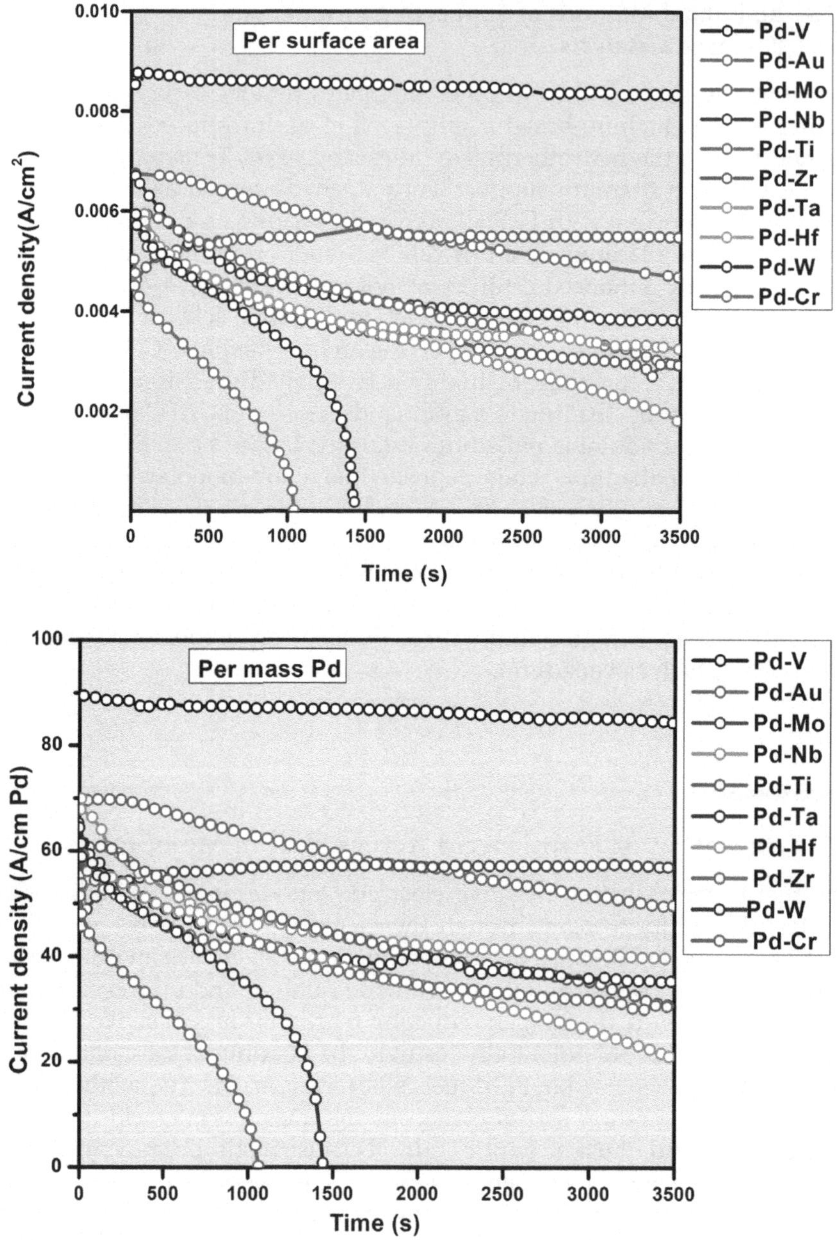

FIGURE 5.10
Chronoamperometric performance of Pd-M catalysts at 0.3 V vs. RHE, as measured in (a) per mass of Pd and (b) as a function of surface area with Pd = 0.6 (M = Ti, Zr, Hf, V, Nb, Ta, Cr, Mo, W, Au). A solution with 5 M HCOOH and 0.1 M H_2SO_4 was used for measuring the I-t graphs [50].

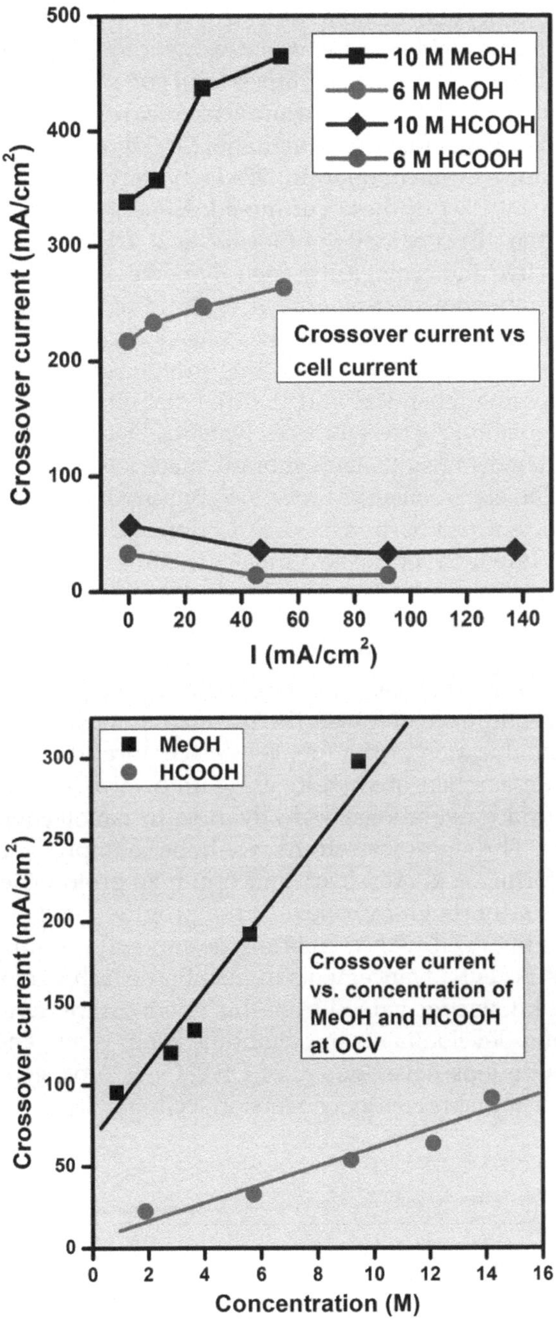

FIGURE 5.11
Methanol and formic acid crossover flow analysis. (a) The crossover current at different concentrations of MeOH and HCOOH compared to the cell current under load; (b) crossover current at different concentrations of MeOH and HCOOH at OCV [53].

reaching the cathode) with comparable cell operating conditions, it was demonstrated that formic acid crossing was nearly six times greater than that of methanol [53]. But when comparing fluxes (mol cm^2 s^{-1}), the difference is just two or less since the oxidation of formic acid only requires two electrons as opposed to six for the oxidation of methanol [54,53].

The examination of membranes in DFAFCs offers insightful information about a significant part of these cutting-edge energy conversion technologies. Through the direct oxidation of formic acid, DFAFCs provide the possibility for effective and ecologically friendly power generation. The DFAFC system's membrane, however, is crucial to deciding how well it functions overall and how long it will last. This extensive review explores the many facets of membranes in DFAFCs, including proton conductivity, preventing formic acid crossing, chemical and thermal stability, mechanical durability, and the interactions between these elements. One cannot exaggerate the importance of choosing suitable membrane materials with excellent proton conductivity. The electrochemical reactions required for energy production depend on efficient proton transport. Additionally, the problem of formic acid crossover, in which formic acid molecules infiltrate the membrane and reduce fuel efficiency, necessitates the development of novel tactics to reduce this phenomenon while preserving proton transport capacities. Given that DFAFCs operate in acidic environments and at a range of temperatures, the chemical and thermal stability of membranes is a crucial component in assuring their endurance. Because the membrane must keep its structural integrity for the duration of the fuel cell, mechanical durability is another crucial factor. Researchers are investigating innovative materials, composite membranes, surface alterations, and advances in nanotechnology to solve these problems. Through these efforts, we hope to improve membrane stability, reduce formic acid crossover, and optimize proton conduction. This paper provides a thorough overview of the progress made in tackling the multiple issues involved with membrane design and optimization by analyzing the most recent advancements in membrane technology for DFAFCs. Ultimately, the knowledge gained from this review serves as a road map for future study and developments in membrane composition and design, promoting the continuous development of DFAFCs as prospective contenders for clean and sustainable energy conversion technologies.

5.4 Conclusions

An overview of recent developments in fuel cells utilizing DFAFCs is given in this article. The underlying chemistry of DFAFCs, anodic catalysts for the oxidation of formic acid and Nafion membranes with formic acid permeability, and modifications to DFAFC topologies are all significant subjects covered. The low standard potential of formic acid causes DFAFCs to have a noticeably high theoretical OCV of 1.48 V. The "dual pathway" oxidation of formic acid frequently involves both a dehydrogenation process that produces CO_2 directly and a dehydration step that creates an intermediate "CO" that is subsequently oxidized to produce gaseous CO_2. Both Pt- and Pd-based catalysts are routinely employed in the synthesis of DFAFCs, with Pd becoming increasingly significant. The PtRu, PtBi, PtPd, PtAu, PtPb, and PtIr are just a few of the Pt-based bimetallic complexes that have been reported to function well as DFAFC catalysts. Although Pd-based catalysts frequently have higher formic acid oxidation activity, they may eventually lose some of that activity. Fortunately, giving the anode a high potential can aid in their regaining some of their earlier activity. Recent years have seen a significant increase in scientific interest in catalysts with carbon support that are based on palladium. Diverse synthetic techniques have produced catalysts with Pd/C or Pd-M/C, which have exceptional DFAFC catalytic activity and a high palladium usage efficiency. The most current quantitative experiments involving formic acid flux across Nafion membranes and crossover behavior are described in this paper. With Nafion membranes, formic acid demonstrates good compatibility. The findings demonstrate that the transitional flux for formic acid is just roughly half as little as the methanol crossover flux. The formic acid crossover's impact on cathode performance, however, is substantially less than that of methanol because there are fewer electrons required to oxidize formic acid (two vs. six). For usage in small portable fuel cells, DFAFCs are typically seen as attractive portable power equipment. A wide variety of formic acid concentrations have been successfully demonstrated for both passive and active micro DFAFCs. DFAFCs are attractive candidates to fulfill rising demands for power density since micro fuel cells have a huge potential to produce better energy density per unit weight and volume than standard batteries. DFAFCs are a potential power source for the foreseeable future due to inherent benefits, including their high electromotive force, minimal fuel crossover, and feasible power densities at low temperatures.

References

[1] Ullah, A., Hussain, A., Qamar, S., Ali, M. M., & Rehman, S. U. (2023). Green materials and methods for wastewater treatment and remediation. In Baskar, C., Ramakrishna, S., & Daniela La Rosa, A. (eds.) *Encyclopedia of green materials*. Springer, Singapore. https://doi.org/10.1007/978-981-16-4921-9_52-1

[2] Ahmad, S., Saher, Dr. S., Ahmed, M., & Ullah, A. (2019). Performance evaluation of nickle platinum novel electrocatalysts for proton exchange membrane fuel. *International Journal of Engineering Works*, 6(01), 39–43.

[3] Ullah, A., Khan, K., Alam, K., Ahsan, I., & Humayun, M. (2018). Synthesis of Ni nanotubes for enhanced oxygen reduction reaction at cathode side of PEM fuel cell. *International Journal of Engineering Works*, 5(10), 188–192.
[4] Ullah, D. M. H., & Holman, Z. Design and fabrication of super-hydrophilic, antifogging and antireflective nanoparticles coatings for solar PV application. https://www.researchgate.net/publication/342178246_design_and_fabrication_of_super-hydrophilic_antifogging_and_antireflective_nanoparticles_coatings_for_solar_pv_application
[5] Ahmad, S., Nawaz, T., Ullah, A., Ahmed, M., Khan, M. O., Saher, S., . . . & Sikandar, M. A. (2021). Thermal optimization of manganese dioxide nanorods with enhanced ORR activity for alkaline membrane fuel cell. *Electrochemical Science Advances*, 1(3), e2000032.
[6] Ahmed, S., Kazmi, W. W., Hussain, A., Khan, M. Z., Bibi, S., Saleem, M., . . . & Khan, M. K. (2023). Facile and low-temperature synthesis approach to fabricate Sm0. 5Sr0. 5CoO3− δ cathode material for solid oxide fuel cell. *Journal of the Korean Ceramic Society*, 60(2), 272–282.
[7] Alam, K., Khan, K. I., Ullah, A., Ullah, A., Ali, S., Ullah, S., . . . & Hussain, S. (2021). Fabrication of superhydrophillic and graded index antireflective double layer coating for solar photovoltaics module using aerosol impact deposition assembly. *Thin Solid Films*, 721, 138518.
[8] Hussain, K., Ali, I., Hasnain, S., Hussain, S. S., Hussain, B., Khan, M. S., . . . & Abbas, N. (2020). Reagents assisted Mg-doped CeO2 for high-performance energy-storage applications. *Journal of Electroanalytical Chemistry*, 873, 114401.
[9] Saher S, Alam K, Qamar A, Ullah A., & Iqbal J. (2019). Influence of dual layer silica nanoparticles coating on the performance enhancement of solar PV modules. *Journal of Mechanics of Continua and Mathematical Sciences*, 365–375.
[10] Shahbaz, A., Afaf, A., Tahir, N., Abid, U., & Saim, S. (2021). Non precious metal catalysts: A fuel cell and ORR study of thermally synthesized nickel and platinum mixed nickel nanotubes for PEMFC. *Key Engineering Materials*, 875, 193–199.
[11] Ullah, A., Hussain, B., & Hussain, S. S. (2021, November). Synthesizing and Optimizing Manganese Dioxide Nanorods and Its Behavior Toward Oxygenreduction Reaction. In 2021 AIChE Annual Meeting. AIChE.
[12] Rice, C., Ha, S., Masel, R. I., Waszczuk, P., Wieckowski, A., & Barnard, T. (2002). Direct formic acid fuel cells. *Journal of Power Sources*, 111(1), 83–89.
[13] Ullah, A., Hussain, S. S., & Hussain, B. (2022, October). Ruthenium Doped Lscf Based Cathode for Enhanced Performance of Solid Oxide Fuel Cells. In Electrochemical Society Meeting Abstracts 242 (No. 23, pp. 965–965). The Electrochemical Society, Inc.
[14] Ullah, A., Hussain, S. S., & Hussain, B. (2022, June). Fabrication of Efficient Mixed Cation Materials for Perovskite Solar Cells to Enhance the Stability and Conversion Efficiency. In 241st ECS Meeting (May 29–June 2, 2022). ECS.
[15] Ullah, A., Hussain, S. S., & Hussain, B. (2022, June). Synthesis of MnO2 Carbon Nanotubes Catalyst with Enhanced Oxygen Reduction Reaction for Polymer Electrolyte Membrane Fuel Cell. In 241st ECS Meeting (May 29–June 2, 2022). ECS.
[16] Ullah, A., Hussain, S. S., & Hussain, B. (2022, May). Fabrication of Efficient Anode By Integrating Transition Bimetallic Oxide with Carbon Nanotubes for Lithium Ion Battery. In 241st ECS Meeting (May 29–June 2, 2022). ECS.

[17] Hussain, B., Ullah, A., Abbas, W., & Hussain, S.S. (2022). Biomass-Derived Materials for Lithium Secondary Batteries. In: Baskar, C., Ramakrishna, S., & Daniela La Rosa, A. (eds) Encyclopedia of Green Materials. Springer, Singapore. https://doi.org/10.1007/978-981-16-4921-9_213-1

[18] Lović, J. D., Elezović, N. R., Jović, B. M., Zabinski, P., Gajić-Krstajić, L., & Jović, V. D. (2018). Electrodeposited AgPd alloy coatings as efficient catalysts for the ethanol oxidation reaction. *International Journal of Hydrogen Energy*, 43(39), 18498–18508.

[19] Capon, A., & Parson, R. (1973). The oxidation of formic acid at noble metal electrodes: I. review of previous work. *Journal of Electroanalytical Chemistry and Interfacial Electrochemistry*, 44(1), 1–7.

[20] Capon, A., & Parsons, R. (1973). The oxidation of formic acid on noble metal electrodes: II. a comparison of the behaviour of pure electrodes. *Journal of Electroanalytical Chemistry and Interfacial Electrochemistry*, 44(2), 239–254.

[21] Weber, M., Wang, J. T., Wasmus, S., & Savinell, R. F. (1996). Formic acid oxidation in a polymer electrolyte fuel cell: A real-time mass-spectrometry study. *Journal of the Electrochemical Society*, 143(7), L158.

[22] Mathias, M., Roth, J., Fleming, J., Lehnert, W., & Vielstich, W. (2003). Handbook of fuel cells—fundamentals, technology and applications. *Fuel Cell Technology and Applications*, 3, 517–537.

[23] Rice, C., Ha, S., Masel, R. I., & Wieckowski, A. (2003). Catalysts for direct formic acid fuel cells. *Journal of Power Sources*, 115(2), 229–235.

[24] Waszczuk, P., Barnard, T. M., Rice, C., Masel, R. I., & Wieckowski, A. (2002). A nanoparticle catalyst with superior activity for electrooxidation of formic acid. *Electrochemistry Communications*, 4(7), 599–603.

[25] Thomas, F. S., & Masel, R. I. (2004). Formic acid decomposition on palladium-coated Pt (1 1 0). *Surface Science*, 573(2), 169–175.

[26] Choi, J. H., Jeong, K. J., Dong, Y., Han, J., Lim, T. H., Lee, J. S., & Sung, Y. E. (2006). Electro-oxidation of methanol and formic acid on PtRu and PtAu for direct liquid fuel cells. *Journal of Power Sources*, 163(1), 71–75.

[27] Casado-Rivera, E., Volpe, D. J., Alden, L., Lind, C., Downie, C., Vázquez-Alvarez, T., . . . & Abruña, H. D. (2004). Electrocatalytic activity of ordered intermetallic phases for fuel cell applications. *Journal of the American Chemical Society*, 126(12), 4043–4049.

[28] Alden, L. R., Han, D. K., Matsumoto, F., Abruña, H. D., & DiSalvo, F. J. (2006). Intermetallic PtPb nanoparticles prepared by sodium naphthalide reduction of metal-organic precursors: Electrocatalytic oxidation of formic acid. *Chemistry of Materials*, 18(23), 5591–5596.

[29] Tripković, A. V., Popović, K. D., Stevanović, R. M., Socha, R., & Kowal, A. (2006). Activity of a PtBi alloy in the electrochemical oxidation of formic acid. *Electrochemistry Communications*, 8(9), 1492–1498.

[30] Matsumoto, F., Roychowdhury, C., DiSalvo, F. J., & Abruña, H. D. (2007). Electrocatalytic activity of ordered intermetallic PtPb nanoparticles prepared by borohydride reduction toward formic acid oxidation. *Journal of the Electrochemical Society*, 155(2), B148.

[31] Zhang, L. J., Wang Z. Y., & Xia, D. G. (2006). Bimetallic PtPb for formic acid electro-oxidation. *Journal of Alloys and Compounds*, 426(1–2), 268–271.

[32] Herrero, E., Fernandez-Vega, A., Feliu, J. M., & Aldaz, A. (1993). Poison formation reaction from formic acid and methanol on Pt (111) electrodes modified by irreversibly adsorbed Bi and As. *Journal of Electroanalytical Chemistry*, 350(1–2), 73–88.

[33] Xia, X. H., & Iwasita, T. (1993). Influence of underpotential deposited lead upon the oxidation of HCOOH in HClO4 at platinum electrodes. *Journal of the Electrochemical Society*, 140(9), 2559.
[34] Arenz, M., Stamenkovic, V., Ross, P. N., & Markovic, N. M. (2004). Surface (electro-) chemistry on Pt (1 1 1) modified by a Pseudomorphic Pd monolayer. *Surface Science*, 573(1), 57–66.
[35] Gojković, S. L., Tripković, A. V., Stevanović, R. M., & Krstajić, N. V. (2007). High activity of Pt4Mo alloy for the electrochemical oxidation of formic acid. *Langmuir*, 23(25), 12760–12764.
[36] Demirci, U. B., Akdim, O., Hannauer, J., Chamoun, R., & Miele, P. (2010). Cobalt, a reactive metal in releasing hydrogen from sodium borohydride by hydrolysis: A short review and a research perspective. *Science China Chemistry*, 53, 1870–1879.
[37] Zhou, X., Xing, W., Liu, C., & Lu, T. (2007). Platinum-macrocycle co-catalyst for electro-oxidation of formic acid. *Electrochemistry Communications*, 9(7), 1469–1473.
[38] Ha, S., Larsen, R., Zhu, Y., & Masel, R. I. (2004). Direct formic acid fuel cells with 600 mA cm–2 at 0.4 V and 22° C. *Fuel Cells*, 4(4), 337–343.
[39] Zhu, Y., Khan, Z., & Masel, R. I. (2005). The behavior of palladium catalysts in direct formic acid fuel cells. *Journal of Power Sources*, 139(1–2), 15–20.
[40] Jung, W. S., Han, J., & Ha, S. (2007). Analysis of palladium-based anode electrode using electrochemical impedance spectra in direct formic acid fuel cells. *Journal of Power Sources*, 173(1), 53–59.
[41] Lović, J. D., Tripković, A. V., Gojković, S. L., Popović, K. D., Tripković, D. V., Olszewski, P., & Kowal, A. (2005). Kinetic study of formic acid oxidation on carbon-supported platinum electrocatalyst. *Journal of Electroanalytical Chemistry*, 581(2), 294–302.
[42] Larsen, R., Ha, S., Zakzeski, J., & Masel, R. I. (2006). Unusually active palladium-based catalysts for the electrooxidation of formic acid. *Journal of Power Sources*, 157(1), 78–84.
[43] Liu, Z., Hong, L., Tham, M. P., Lim, T. H., & Jiang, H. (2006). Nanostructured Pt/C and Pd/C catalysts for direct formic acid fuel cells. *Journal of Power Sources*, 161(2), 831–835.
[44] Zhang, L., Tang, Y., Bao, J., Lu, T., & Li, C. (2006). A carbon-supported Pd-P catalyst as the anodic catalyst in a direct formic acid fuel cell. *Journal of Power Sources*, 162(1), 177–179.
[45] Zhang, L., Lu, T., Bao, J., Tang, Y., & Li, C. (2006). Preparation method of an ultrafine carbon supported Pd catalyst as an anodic catalyst in a direct formic acid fuel cell. *Electrochemistry Communications*, 8(10), 1625–1627.
[46] Li, X., & Hsing, I. M. (2006). Electrooxidation of formic acid on carbon supported PtxPd1– x (x= 0–1) nanocatalysts. *Electrochimica Acta*, 51(17), 3477–3483.
[47] Uhm, S., Chung, S. T., & Lee, J. (2007). Activity of Pt anode catalyst modified by underpotential deposited Pb in a direct formic acid fuel cell. *Electrochemistry Communications*, 9(8), 2027–2031.
[48] Park, I. S., Lee, K. S., Choi, J. H., Park, H. Y., & Sung, Y. E. (2007). Surface structure of Pt-modified Au nanoparticles and electrocatalytic activity in formic acid electro-oxidation. *The Journal of Physical Chemistry C*, 111(51), 19126–19133.

[49] Yi, Q., Chen, A., Huang, W., Zhang, J., Liu, X., Xu, G., & Zhou, Z. (2007). Titanium-supported nanoporous bimetallic Pt–Ir electrocatalysts for formic acid oxidation. *Electrochemistry Communications*, 9(7), 1513–1518.
[50] Larsen, R., Zakzeski, J., & Masel, R. I. (2005). Unexpected activity of palladium on vanadia catalysts for formic acid electro-oxidation. *Electrochemical and Solid-State Letters*, 8(6), A291.
[51] Heinzel, A., & Barragan, V. M. (1999). A review of the state-of-the-art of the methanol crossover in direct methanol fuel cells. *Journal of Power Sources*, 84(1), 70–74.
[52] McGrath, K. M., Prakash, G. S., & Olah, G. A. (2004). Direct methanol fuel cells. *Journal of Industrial and Engineering Chemistry*, 10(7), 1063–1080.
[53] Jeong, K. J., Miesse, C. M., Choi, J. H., Lee, J., Han, J., Yoon, S. P., . . . & Lee, T. G. (2007). Fuel crossover in direct formic acid fuel cells. *Journal of Power Sources*, 168(1), 119–125.
[54] Wang, X., Hu, J. M., & Hsing, I. M. (2004). Electrochemical investigation of formic acid electro-oxidation and its crossover through a Nafion® membrane. *Journal of Electroanalytical Chemistry*, 562(1), 73–80.

6

Advances in Membrane and Catalyst in Direct Dimethyl Ether Fuel Cells (DDMEFCs)

Siti Hasanah Osman, Siti Kartom Kamarudin, and Norazuwana Shaari

6.1 Introduction

Particularly in the domain of portable technology encompassing devices such as laptops, mobile phones, personal digital assistants (PDAs), and personal media players (PMPs), the demand for batteries with greater power capacity is growing. Despite extensive investigation, lithium-ion and metal hydride batteries still have certain drawbacks in terms of augmenting power output without introducing additional bulk to the apparatus. The utilization of alternative energy sources is critical. Fuel cells operate at elevated temperatures and pressures due to the direct feeding principle they employ; this characteristic has established them as a potentially viable alternative to traditional batteries in portable applications [1]. Fuel cell variants that operate on liquid fuels (e.g., dimethyl ether, methanol, and formic acid) have brought about a paradigm shift in portable energy alternatives. Fuel cell research has garnered the attention of academic and industrial research institutions alike. Direct methanol fuel cell (DMFC) technology is approaching commercial viability and is the most advanced of these technologies. Conversely, substantial impediments that inhibit large-scale manufacturing impede the application of methanol as a fuel for DMFCs. Methanol presents an environmental hazard owing to its propensity for infiltration into groundwater and its deleterious impacts on the respiratory and cutaneous systems. Other liquid fuels, including ethanol and formic acid, frequently demonstrate substandard performance or significant corrosion problems. In contrast, dimethyl ether (DME) emerges as a fuel that circumvents these limitations and possesses some noteworthy benefits, as was previously stated [2–4]. Among the numerous notable benefits of DME are the following: Theoretical

DOI: 10.1201/9781003512592-6

fuel consumption is minimized by the high electron transfer number of 12 in DME, which is necessary for complete oxidation to CO_2. Methanol, in contrast, consists of only two hydrogen atoms and six electrons. It is feasible to achieve complete oxidation of DME without any loss of kinetic energy, as it is the simplest ether devoid of C-C bonds. DME diminishes the dipole moment of DME as it transitions from the anode to the cathode, in contrast to methanol. Transferring DME from a tank to the fuel cell system is straightforward and does not require an additional compressor; the process is analogous to storing LPG under high pressure. DME is less explosive, less poisonous, and does not leach into groundwater in the same way that methanol and hydrogen do.

DME is less active than methanol under these conditions, although the efficiency of DDMEFCs is comparable to that of conventional catalysts (at 50°C and 1 bar pressure) and DMFCs operating under normal conditions (at 130°C and 1 bar pressure). Thus, dimethyl ether exhibits considerable potential as a direct-injection fuel cell propellant. Researchers from around the globe have devoted the last decade to studying the mechanisms underlying DME oxidation to create novel electrocatalysts. Through catalytic conversion with syngas, DME can be converted into various useful compounds, including formaldehyde [5], methyl acetate [6], and ethanol [7]. DME finds additional applications as a solvent, kiln fuel for ceramics, and in the context of welding, cutting, and brazing. On the subject of DME synthesis, numerous scholarly articles have been published over the years (**see** Figure 6.1(a)), especially in the last two decades. In 2019, the greatest number of articles were published on the subject because of its perceived great potential as a fuel, which drew the interest of academic and industrial researchers. Therefore, it is opportune to reassess the extent of DME's utility as a precursor and fuel for various chemical reactions, as well as in industrial processes. Figure 6.1(b) suggests that the articles under consideration place a considerable focus on

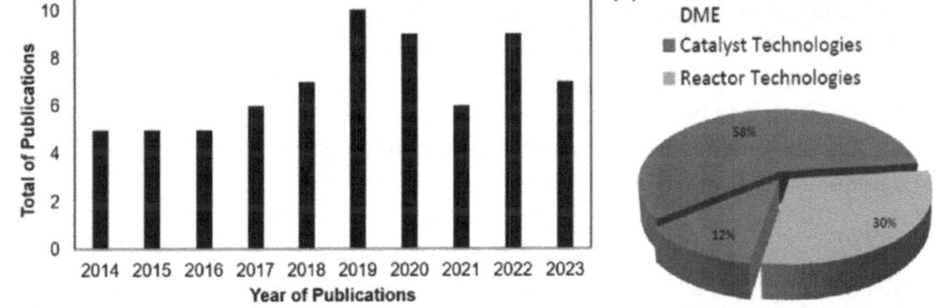

FIGURE 6.1
The trend of related publications on DME.

catalyst technology prior to exploring reactor technology. In light of these findings, this research focuses on production processes. In addition to talking about their wide range of reactors and catalyst setups, they also look at effective aspects, including water removal, the H2/CO ratio, the feed's CO_2 level, temperature, pressure, and space velocity.

Furthermore, almost all the findings of the study demonstrated that carbon-based materials are useful in DME applications focusing on electrocatalysts and MEA, resulting in strong performance. Besides, Serov [8], Mondal [9], and Azizi [10] presented a review paper focusing on the development of DDMEFCs. However, this work presents a wide range of carbon-based applications on DDMEFCs, including electrocatalysts and MEA. In an attempt to improve the efficiency of DDMEFCs, optimization of MEA performance, modification of the fuel mixture, and investigation of various operating conditions have all been implemented. The current state of research in DDMEFCs is the subject of this article.

6.2 Mechanism of Dimethyl Ether Oxidation

Several potential oxidation pathways for dimethyl ether (DME) have been discerned via an analysis of byproducts from an anode exhaust. According to their hypothesis, a solitary methyl group undergoes oxidation and hydrolysis, resulting in the formation of MeOH via the cleavage of the C-O-C bond, as suggested by these results. The operational mechanisms at the anode end of DME fuel cells have been the focus of numerous studies in recent times [11–13]. The findings of Mizutani et al. [14] regarding the oxidation of DME on Pt/C catalysts demonstrated that the efficiency of the CO2 current increased with temperature in half-cell experiments. Electro-oxidation of methanol is consistent with the temperature-dependent increase in CO_2 current efficiency that has been observed [15]. Upon comparing these values, it becomes evident that DME exhibits a superior current efficacy in comparison to methanol [16].

Additionally, the authors investigated the byproducts of the reactions occurring at the anode of the membrane electrode assembly (MEA) within a DDMEFC. It has been demonstrated that DME predominantly oxidizes to CO_2 at high temperatures, which is consistent with the outcomes of half-cell tests. DME oxidation in DDMEFC generated methanol and methyl formate as the principal byproducts. The probability that methanol could be generated via an electrochemical mechanism is thus minimal. A potential synthesis process, as described in Equation (6.1), is H_2O hydrolysis.

$$CH_3OCH_3 + H_2O \rightarrow 2CH_3OH \tag{6.1}$$

The electrochemical nature of this reaction is demonstrated by the proportional increase in methyl formate production rate as current density increases. The methyl formate is likely generated via the procedure described in Equation (6.2):

$$CH_3OCH_3 + H_2O \rightarrow HCOOCH_3 + 4H^+ + 4e^- \quad (6.2)$$

The temperature-dependent examination of methyl formate production leads the authors to conclude that the mechanism of generating CO_2 is preferable to that of generating $HCOOCH_3$. Examining the electro-oxidation of DME on a platinum electrode in bulk, Kerangueven and team [17] found that DME adsorption is a progressive step within the overall oxidation process. They also pointed out that DME adsorption can depend on the platinum's hydrogen area but not on the double-layer region. DME extraction and low potential scan rate voltammetry experiments showed that there are several chemical pathways involved in DME oxidation. In situ IR reflectance spectroscopy was used to detect CO_2 in addition to adsorbed COOH species, linearly bound CO (COL), and bridge-bonded CO (COB). These electrochemical and spectro-electrochemical results led the authors to conjecture that part of the adsorbed DME was directly hydrolyzed or oxidized to CO_2 or HCOOH species, while Pt-CHO and/or PtCO surface species were maintained at the platinum sites. Drawing from this thorough analysis, the authors hypothesized the following process of DME oxidation: at low potentials, DME adsorption blocks a portion of the platinum surface, allowing direct DME oxidation to take place on the remaining platinum surface. This mechanism needs to take into account the synthesis of methanol, formic acid, and CO_2 at low potentials, as well as linearly and bridge-bonded CO species. A current plateau could arise from DME's direct oxidation in the low potential range due to molecular breakdown and C-adsorption:

$$CH_3OCH_3 + H_2O \rightarrow CH_2OH_{ad} + H^+ + e^- \quad (6.3)$$

and/or

$$CH_3OCH_3 + H_2O \rightarrow CH_2OH_{ad} + 2H^+ + 2e^- \quad (6.4)$$

followed by

$$CH_2OH_{ad} + CHO_{ad} \rightarrow 2H^+ + 2e^- \quad (6.5)$$

then

$$CHO_{ad} + H_2O \rightarrow CO_2 + 3H^+ + 3e^- \quad (6.6)$$

and/or

$$CHO_{ad} + H_2O \rightarrow HCOOH + H^+ + e^- \quad (6.7)$$

and/or

$$CHO_{ad} + CO_{ad} \rightarrow H^+ + e^- \quad (6.8)$$

At higher potentials (higher than 0.55 V vs. RHE), platinum is able to activate water and can adsorb OH species:

$$CHO_{ad} + OH_{ad} \rightarrow CO_2 + 2H^+ + 2e^- \quad (6.9)$$

and

$$CHO_{ad} + OH_{ad} \rightarrow CO_2 + H^+ + e^- \quad (6.10)$$

To precisely identify adsorbent intermediates and reaction products, the authors emphasized the importance of conducting more exhaustive in situ IR spectroscopic investigations and examining the correlation between electrode voltage and reaction product distribution.

Liu et al. [18] subsequently investigated dimethyl ether oxidation on a platinum surface while determining the intermediate adsorption species and oxidation pathways of DME. The researchers employed in situ attenuated total reflection-Fourier transform infrared spectroscopy (ATR-FTIR) and electrochemical measurements to ascertain that the DME oxidation reaction at potentials ranging from 0.1 to 0.45 V covered nearly 90% of the adsorption species (uads). The primary intermediate adsorption species on the surface of the catalyst were -COOH, -CHO, -HCOO-, and -OCH_3, excluding CO. In conclusion, they suggested the development of novel electrocatalysts, placing particular emphasis on those that possess sufficient stability to circumvent the detrimental consequences of carbon monoxide. Transition metal-containing platinum alloys (e.g., PtRu and PtSn) might serve as the foundation for these novel catalysts. In their investigation of the mechanism of DME oxidation, Zhang et al. [19] conducted electrochemical and in situ infrared experiments. For dealing with the challenges posed by soluble species' diffusion and accumulation, a creative approach was utilized to design and implement a spectroscopic flow cell equipped with a Kretschmann ATR apparatus for in situ infrared measurements. This approach facilitated the examination of the species that were present on the surface of the electrode through the elimination of soluble residues and the consistent provision of reactant concentration.

6.3 Synthesis Methods

DME can be created by two different routes shown in Figure 6.2. the indirect route uses produced methanol to aid in its dehydration, but the direct route is perhaps more efficient because it uses bifunctional catalysts to produce DME in a single step. Leading companies in the single-step technique technology include Haldor Topsoe, JFE Holdings, Korea Gas Corporation, Air Products, and NKK [20,21]. Additionally, each of Toyo, MGC, Lurgi, and Udhe has developed their indirect procedures for the production of DMW [20]. Syngas, a mixture of hydrogen and carbon monoxide, is produced and is a crucial step in the synthesis of DME. Syngas is produced industrially by converting hydrocarbon fuels, primarily natural gas, using techniques including gasification and steam reforming (SR) [22,23]. The next sections will cover important information about these two fundamental methods of producing DME synthesis and then look into other ways of producing DME.

6.3.1 Direct Synthesis Method

Recent advancements have enabled the integration of the methanol synthesis and dehydration processes [24], enabling the direct synthesis of DME from syngas in a single reactor. The water-gas shift reaction is incorporated into Equation (6.11), which governs the direct synthesis of DME from syngas comprising H_2, CO, and CO_2. Equation (6.12) does not include this reaction.

$$3CO + 3H_2 \rightarrow CH_3OCH_3 + CO_2 \tag{6.11}$$

$$2CO + 4H_2 \rightarrow CH_3OCH_3 + H_2O \tag{6.12}$$

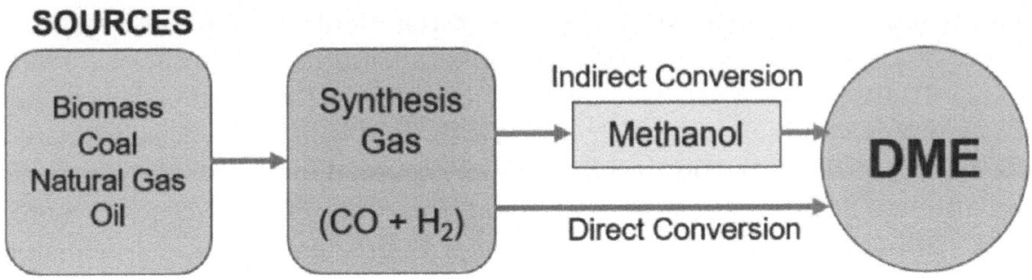

FIGURE 6.2
Dimethyl ether production diagram.

The Equation (6.11) involves four basic reactions, including the following [25]:

Methanol synthesis from CO:

$$CO + 2H_2 \leftrightarrow CH_3OH \qquad \Delta H^{+}_{298K} = -90.4\ kJ/mol \tag{6.13}$$

Methanol synthesis from CO_2:

$$CO_2 + 3H_2 \leftrightarrow CH_3OH + H_2O \qquad \Delta H^{+}_{298K} = -49.4\ kJ/mol \tag{6.14}$$

Water gas shift (WGS):

$$CO + H_2O \leftrightarrow CO_2 + H_2 \qquad \Delta H^{+}_{298K} = -41.0\ kJ/mol \tag{6.15}$$

Methanol dehydration:

$$2CH_3OH \leftrightarrow CH_3OCH_3 + H_2O \qquad \Delta H^{+}_{298K} = -23.0\ kJ/mol \tag{6.16}$$

The maximal conversion of each reaction occurs when the H_2/CO ratio in the feed stream matches the stoichiometric value. For Equation (6.11), this value is 1.0; for Equations (6.12) and (6.13), it is 2.0. The aforementioned reactions demonstrate that the syngas to the DME (STD) pathway comprises methanol synthesis and WGS reactions. The former is an essential component in the synthesis of DME, while the latter generates CO_2, which serves as the primary byproduct [26]. Syngas can be generated with an H_2/CO molar ratio of 1 by utilizing CO_2 in a methane reforming unit by the reaction stoichiometry of Equation (6.17).

$$2CH_4 + O_2 + CO_2 \rightarrow 3CO + 3H_2 + H_2O \tag{6.17}$$

Accurate temperature regulation is imperative to avert catastrophic reactions, given that the STD process is fundamentally a highly exothermic reaction. Although it reduces the emissions of pollutants from natural gas, the direct synthesis method is among the most challenging for methane conversion. In the synthesis process, chemical engineering operating devices are frequently used to separate and purify DME. H_2, N_2, and CO_2 are eliminated via distillation, absorption, and flash [27]. Recovered methanol is acquired as the ultimate DME product. As a thermodynamically limited process, methanol synthesis shifts its equilibrium towards greater methanol conversion when methanol is utilized in the succeeding reaction to produce DME [10]. CO_2 and DME separation are complicated by the presence of methanol. As recommended, the methanol and water produced by a one-step reaction are condensed and absorbed by water. In essence, the DME product is extracted from the liquid stream containing DME via distillation [28]. Developing DME

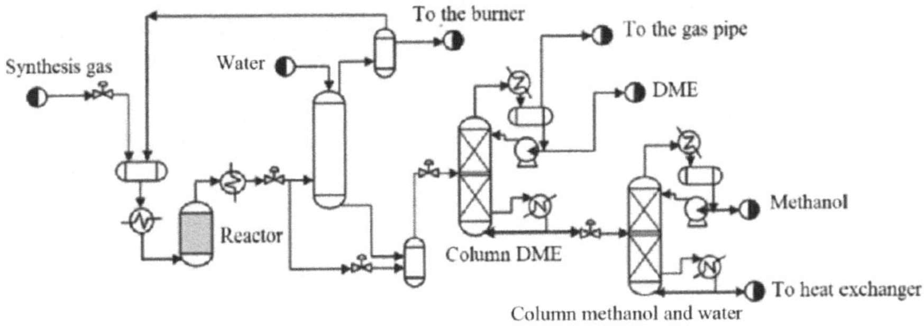

FIGURE 6.3
A scheme of direct synthesis process.

of exceptional purity has been demonstrated to be feasible using this separation method. The schematic illustrating this process is depicted in Figure 6.3.

6.3.2 Indirect Synthesis Method

The conventional method for producing DME entailed a two-step sequence beginning with the production of methanol from syngas, followed by its purification and conversion into DME in a second reactor. Figure 6.4 illustrates this procedure schematically. The dehydration of methanol to produce DME is represented by Equation (6.18) in the context of commercial processes.

$$2CH_3OH \rightarrow CH_3OCH_3 + H_2O \; \Delta H^{+}_{298K} = -23.5 \text{ kJ/mol} \tag{6.18}$$

Several scholarly publications have been devoted to the kinetics of DME synthesis via solid-acid catalyst dehydration of methanol. The prevailing consensus among scholars is that the mechanism described suits either the Langmuir-Hinshelwood [29] or Eley-Rideal kinetic model [30], wherein water and DME function as inhibitors of the reaction [31].

Bondiera and Naccache [32] provide the rate equation for temperatures exceeding 250°C that corresponds to Equation (19):

$$-r_{methanol} = k_0 \, exp + \left(-\frac{E_a}{RT} \right) P_{methanol} \tag{19}$$

The equations that describe the methanol pressure in kPa (k_0 = 1.21 × 106) and the methanol partial pressure in kPa ($p_{methanol}$) are as follows: 80.48 kJ/mol is Ea. Theoretically, dehydrating methanol at lower temperatures is more advantageous due to the exothermic nature of the reaction and the production of carbon monoxide, ethylene, hydrogen, and charcoal in substantial quantities at higher temperatures.

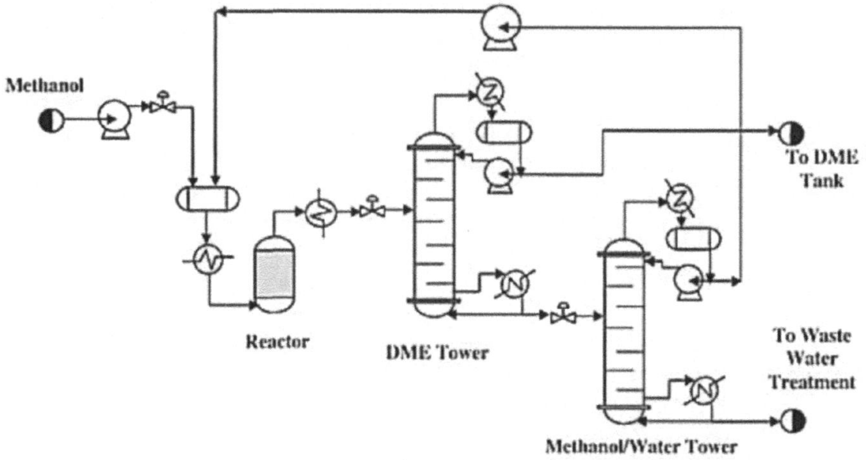

FIGURE 6.4
A Scheme of indirect synthesis process.

6.3.3 Comparison of Methods and Seeking Other Routes

The direct method exhibits superiority over the methanol dehydration method in the production of DME due to its ability to achieve a greater CO conversion and its streamlined reactor architecture [33]. This results in a substantial reduction in the expenses associated with the production of DME. The unreacted syngas and CO_2 produced during the one-step synthesis process, on the other hand, make the separation procedure for high-purity DME comparatively more challenging. The direct procedure is considered preferable to the methanol dehydration method for the production of DME due to its ability to achieve a higher CO conversion and its simpler reactor architecture [33]. As a result, the expenses associated with the production of DME are substantially reduced. Due to the production of unreacted syngas and CO_2 during the one-step synthesis process, the separation procedure for high-purity DME is marginally more complicated.

Furthermore, the commercial viability of directly synthesizing DME from syngas is jeopardized by the water-gas shift reaction, which requires a stoichiometric amount of CO to produce hydrogen and CO_2. Creating DME from syngas is typically a greenhouse gas-emitting and energy-intensive procedure. There is a significant demand for design solutions that are both energy-efficient and environmentally sustainable [34]. These methods encompass both recycling and the conversion of the CO_2 byproduct into the DME synthesis line. Considerable academic attention has been devoted to the utilization of carbon dioxide (CO_2) as a precursor in the fabrication of liquid energy carriers, as it is regarded as a potential solution to the accumulation of CO_2 in the atmosphere [35]. The conventional process has not been granted industrial authorization as a result of its inadequate selectivity, poor DME yield, and low CO_2 conversion efficiency. One viable and efficient strategy to overcome the thermodynamic constraints of DME synthesis is to integrate the selective elimination of H_2O into the equilibrium-limited procedure. This would enable a higher rate of CO_2 conversion [36].

A novel approach is introduced for the synthesis of DME from methane using a two-step procedure: initially, methane is subjected to elevated temperatures, while hydrogen halide and oxygen halide pass over a Rh-SiO_2 catalyst. This results in the formation of methyl halide, typically CH_3Cl or CH_3Br. Methyl halide is hydrolyzed to DME in the second stage using a silica-supported metal chloride catalyst [34,37]. Methanol is the primary product of this process; conversely, corrosion is the primary concern when organic halides are hydrolyzed. Surya Prakash et al. [36] investigated the hydrolysis of methyl bromide in a batch reactor using PVP as a novel catalyst. By utilizing the solid PVP as a potential reusable solid amine catalyst, the maximal effectiveness of the catalyst was determined to be achieved, as indicated by its capacity to trap HBr. The reusability and renewal of the polymer without compromising its functionality is a significant benefit of this method. Dimethyl ether can also be generated through the oxidative carbonylation of methyl bromide. As a catalyst for this procedure, metal oxide or SbF5/graphite could be utilized. Methyl acetate is generated as a byproduct in the initial instance, whereas methyl alcohol is acquired in the subsequent case [38,39].

6.4 Maximizing DDMEFCs Efficiency Performance

The design of MEAs, DME crossover, electrocatalysts, and operating circumstances are all factors that affect DDMEFC performance. PtRu and other electrocatalysts have a significant level of efficiency. MEA construction affects mass transfer and reaction kinetics. Hydrophobic and hydrophilic coatings must be applied to the anode GDL in order to properly handle DME crossover [40,41]. It is essential to adjust operating parameters, such as temperature and pressure, for maximum efficiency. Enhancing the effectiveness of DDMEFCs necessitates a thorough strategy that considers each of these elements.

6.4.1 Electrocatalysts

DDMEFC anode catalyst enhancement potential was uncovered through an examination of the mechanisms underlying DME oxidation. As demonstrated in the preceding chapter, CO is an intermediate product of dimethyl ether oxidation. Platinum materials, which are frequently employed in fuel cell applications, are recognized for their susceptibility to carbon monoxide (CO) damage. By employing DMFC anode catalysts composed of Pt-M alloys (where M represents Ru, Sn, W, and Mo), both catalytic activity and resistance to the deleterious effects of CO were considerably enhanced [42–44]. Bimetallic platinum-based alloys were analyzed for their activity as catalysts for DME oxidation using an identical methodology. The investigation conducted by Liu et al. [45] regarding Pt-M alloyed catalysts for dimethyl ether oxidation (DME solution in sulfuric acid) was the most exhaustive. A comparison was made between a platinum electrocatalyst and a variety of carbon-supported catalysts (PtRu/C, Pt_3Sn/C, PtW/C, Pt_3Mo/C, PtCo/C, PtNi/C, and Pt_2Cr/C) that possess distinct Pt:M ratios. At potentials below 0.55 V, it is indisputable that PtRu/C and Pt_3Sn/C exhibit greater catalytic activity than Pt/C and other PtM/C catalysts. When E is greater than 0.6 V, Pt/C is more efficient than PtRu/C. This phenomenon may be accounted for by the enhanced oxidation efficiency of harmful species generated on a platinum surface that is exposed during the conversion of DME to CO_2 at higher overpotentials.

Chronoamperometry data on catalyst resistance to hazardous intermediate products during DME oxidation show that CO poisoning results in a significant drop in mass activity at 0.4 V for all catalysts in less than 30 seconds. A reduction in the stabilized current density at 180 s causes the following series of events to occur: More than PtRu/C, Pt_3Sn/C, Pt_3Mo/C, Pt_2Cr/C, PtNi/C, PtCo/C, and PtW/C are Pt/C. In comparison to Pt/C, the DME oxidation reaction (DOR) is accelerated by a ratio of 12 in PtRu/C and by 4.4 in Pt_3Sn/C. Pt_3Mo/C, Pt_2Cr/C, PtNi/C, PtCo/C, and PtW/C have factors that range from 3.4 to 1.3, in contrast to Pt/C [45]. The apparent activation energy falls in the following manner, using the slope of the Arrhenius plots as a basis: Pt/C weighs 46 kJ mol^{-1} compared to Pt_3Sn/C 57 kJ mol^{-1}. According to the previously described results, Pt is more resistant to toxic species when a secondary metal is added during DOR. In the low overpotential region (0.55 V, 50 8C), the PtRu/C electrocatalyst exhibits the strongest resistance to toxic species and the highest electrocatalytic activity when compared to the other binary electrocatalysts. At potentials higher than roughly 0.55 V (50 8C), Pt/C instead outperforms PtRu/C.

An investigation was conducted by Coutanceau et al. [46] into the electrical efficacy of fuel cells that were powered by a mixture of methanol and DME. PtRu, which was initially developed and optimized for MeOH oxidation, has been demonstrated to be the most effective catalyst for DME electrooxidation, according to the reviewed studies. One of the study's numerous recommendations is that it is critical to optimize PtRu for DME oxidation or to develop new materials. An additional approach to enhancing the performance of DDMEFCs is to investigate the impact of support on the catalytic activity of platinum-based materials. PtRu/CNT is a potential anode catalyst for the oxidation of DME.

6.4.2 MEA Fabrication and DME Crossover

The MEA production procedure exerted a substantial impact on the ultimate performance of the fuel cell. In consideration of the water management, fuel cell operations, anode and cathode processes, and the selection of microporous layers (MPLs), gas diffusion layers (GDLs), and membrane varieties, it is possible to optimize performance by evaluating these variables. Numerous endeavors to approximate the impacts of the aforementioned parameters have been made in recent times [11,14,47]. The efficacy of MEAs was examined by Mench et al. [47] both with and without membrane thickness. The precipitous performance decline observed in Nafion 117 (with a thickness of 178 mm) can be attributed to the anode dehydrating to an excessive degree. On the cathode side, thin Nafion 112 (thickness 51 mm) is not advised due to substantial DME or DME intermediate crossover. The process of water transfer from the cathode to the anode is facilitated by the 127 mm thick Nafion 115 [47].

Mizutani et al. [14] conducted an exhaustive investigation into the DME crossover that occurs during MEA operation. A comprehensive analysis of the gas streams from the anode and cathode was performed using gas chromatography (GC) to determine the nature and quantity of fully humidified DME that traversed. The MEA was produced using a Nafion 112 membrane. The exhaust rates of DME and CO_2 are graphed against the current density at temperatures of 30, 50, and 80°C. The graphic effectively demonstrates that open-circuit voltage (OCV) conditions do not preclude the presence of detectable amounts of DME. This discovery provides further evidence that DME can traverse a thin Nafion 112 membrane with relative ease, as hypothesized by Mench and his team [47]. An increase in temperature leads to a reduction in DME crossover. The reason for this is that as temperatures rise, the solubility of DME in water decreases proportionally. Additionally, the increase in water vapor pressure caused by rising temperatures probably contributes to a reduction in the quantity of DME gas present in the anode stream. A reduction in DME crossover occurs as a result of the increased current density, which enhances the utilization of the DME on the anode surface and consequently diminishes the crossover quantity. Moreover, the cathode's water serves as an impediment to the DME gas's permeability, thereby diminishing crossover even further with increasing current density. Thus, to optimize the performance of DDMEFCs, it is suggested that a narrower membrane be utilized at higher temperatures and a thicker membrane be utilized at lower temperatures.

On the surface of the anode GDL, Cai et al. [48] performed an exhaustive analysis of hydrophobic and hydrophilic substances. A comparison was made between membrane electrode assemblies (MEAs) constructed with hydrophilic, hydrophobic, and a novel diffusion layer (PTFE comprising 50% hydrophilic Nafion 1 and 50% hydrophobicNafion 1) on a Nafion 115 membrane using their findings. Figure 6.5 depicts scanning electron microscopy

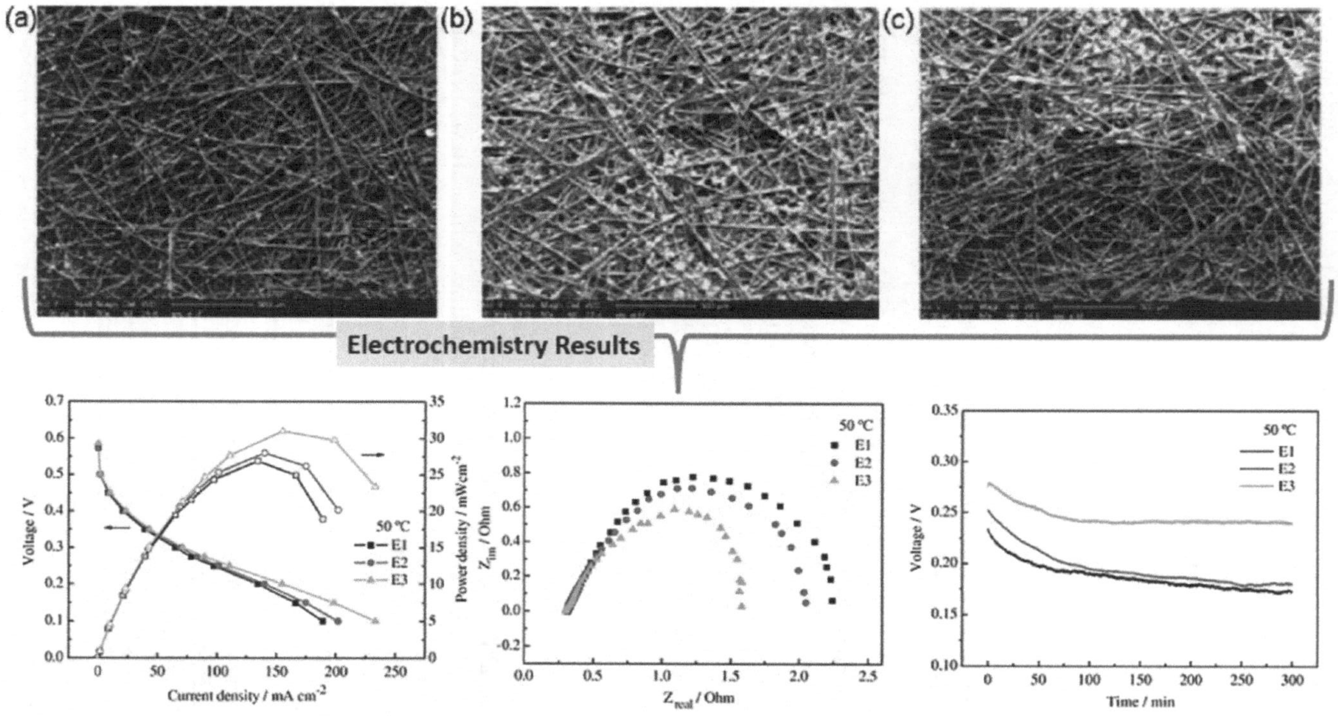

FIGURE 6.5
Three layers of MEA are seen in the SEM pictures of carbon paper backings for anodes: (a) the hydrophilic layer, (b) the hydrophobic layer, and (c) the new layer.

Source: For additional understanding, electrochemistry results are also provided.

images of the anode carbon paper utilized in MEAs (Figures 6.5a and b). Carbon fibers are distributed uniformly across the surface of carbon paper. The hydrophilic and hydrophobic layers are distributed uniformly across the GDL surface, as illustrated in Figure 6.5c. The hydrophobic layer (facilitated by these layers) and the hydrophilic layer (facilitated by water management) collaborate to enhance DME gas control and water management, respectively. Saturated DME solution was utilized in the fuel cell research.

6.4.3 DDMEFCs Operation Conditions

Critical importance is placed on reaction temperature in reversible exothermic processes, such as the synthesis of DME from syngas or a mixture of CO_2 and H_2. Kinetics initially governs the reaction at lesser temperatures. The rate of the reaction, however, is discovered to increase with the temperature of the reactor [49]. Elevated reactor temperature results from the reaction's pursuit of equilibrium, leading to a reduction in the overall conversion rate. Consequently, under such conditions, the temperature of any procedure involving a reversible exothermic reaction must be regulated to decrease gradually. Kumar and Srivastava [50] investigated 230 and 250°C so as to determine the effect of temperature on the synthesis of DME from syngas in a slurry-phase reactor. Aside from a minor decrease in output as the reaction concluded, conversion rates remained constant at both temperatures. Notably, the reaction was more selective at 250°C than at 230°C in the majority of the reactors. Lu et al. [51] have observed that the decrease in activity of the DME synthesis catalyst could potentially be attributed to the sintering of Cu metallic sites. Numerous researchers have substantiated this finding [52,53]. In light of the finding that the methanol dehydration catalyst exhibits greater activity at elevated temperatures than the methanol synthesis catalyst, Kumar and Srivastava [50] elucidated that temperature increases the degree of selectivity. In slurry-phase reactors, the optimal reaction temperature for DME synthesis was generally acknowledged to be 250°C. DME was synthesized from syngas using a bifunctional catalyst comprising CuO-ZnO-MnO/SAPO-18 within a fixed bed reactor functioning at 250–350°C. The paraffin yield, particularly methane, was greatest at temperatures exceeding 300°C, whereas the optimal DME yield was observed at approximately 275°C. The phase transition in DME selectivity occurred between 250 and 275°C, shifting it from 95% to 65% [54]. In their study, Khosbin and Haghighi [55] investigated the synthesis of DME from CO at 200–300°C. They discovered that the conversion of CO reached its highest point at 275°C, after which it commenced to decline at 300°C. This decline in conversion temperature was found to be a contributing factor to the diminished selectivity of DME (in Figure 6.6). Research by Zhau et al. [56] indicates that by increasing the temperature from 240 to 260°C, CO_2 conversion and DME selectivity were enhanced by 21% and 63.1%, respectively. Conversion and selectivity, nevertheless, decreased at approximately 280°C. At elevated

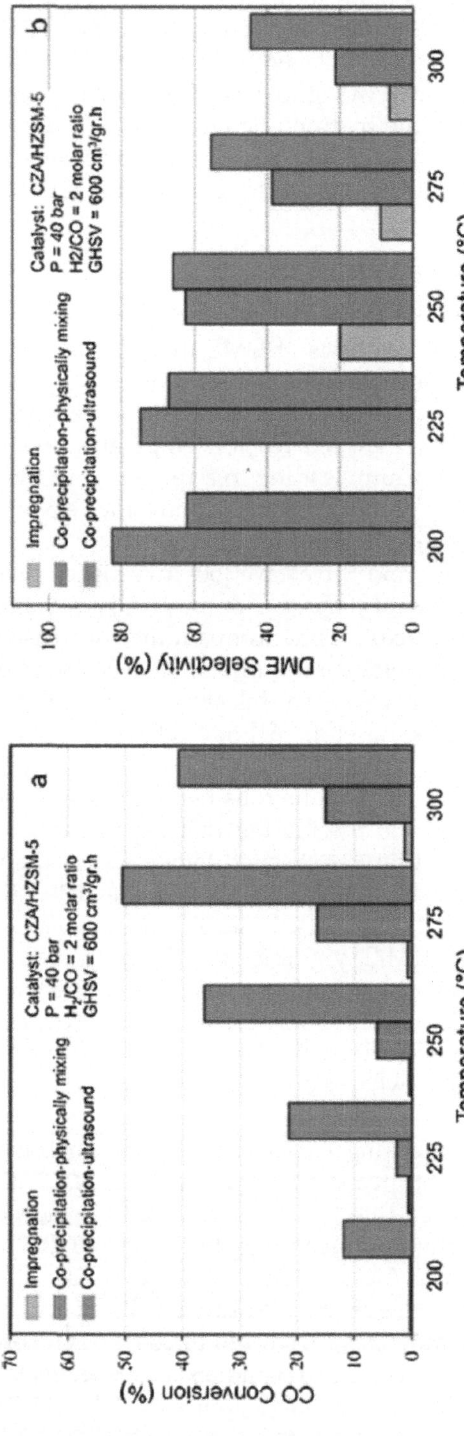

FIGURE 6.6
Variable preparation methods and reaction conditions affect the effect of temperature on CO conversion and DME selectivity using CZA/HZSM-5 catalyst [55].

temperatures, the reversible hydrogenation of CO_2 was observed to result in a reduction in DME selectivity. Similar findings were independently reported by Liu et al. [57] and Zha et al. [38], validating the temperature sensitivity of DME synthesis.

6.5 Conclusions and Future Perspectives

Considerable research effort has been devoted to examining diverse DME production techniques with the aim of enhancing efficiency. Although the prevailing method for producing DME is through the dehydration of pure methanol [58], numerous researchers are investigating the possibility of producing STD directly from syngas using dual heterogeneous catalysts (bifunctional catalysts) in a single reactor. This approach facilitates the concurrent occurrence of methanol synthesis and dehydration. Scientists have demonstrated considerable interest in the application of CZA, while the results indicate that Al_2O_3 and ZSM-5 zeolite are the most frequently used catalysts for methanol dehydration.

Another implication that can be drawn is that researchers have a strong interest in implementing fixed bed reactors due to their ease of operation. In addition, they necessitate substantial capital expenditures for multiple units and have commensurate energy demands. To achieve process intensification, it is imperative to devise process options that are significantly superior to the existing state-of-the-art. Examples of this include the development of catalytic distillation (CD), reactive dividing-wall columns (R-DWC), and dividing-wall columns (DWC) that are considerably more energy-efficient, environmentally friendly, and compact. On the contrary, industrial enterprises require a holistic strategy that facilitates the transition from the conceptual phase to the practical implementation domain. To enhance process performance, the acidity of the sites must guide the formulation of the optimal bifunctional catalyst. Despite this, the topic of catalyst deactivation has not been neglected; acid site covering and pore obstruction mechanisms are hypothesized to be responsible for the inactivation of zeolites, whereas water is hypothesized to influence the deactivation of γ-Al_2O_3. A discernible deactivation of Cu-based catalysts was observed in slurry reactors as opposed to fixed-bed reactors when the two types of reactors were compared.

Additionally, prominent determinants that impact the efficiency of DME production were investigated. Inhibiting chemical reactions is the result of water molecules competing for acid sites with methanol molecules. By displacing the water-gas shift equilibrium, hydrophilic membranes that remove H_2O enhance the process of CO_2 to methanol conversion. The yield and selectivity of DME are thus improved. The influence of temperature on the H_2/CO ratio during DME synthesis is demonstrated to have an optimal value. Higher CO_2 concentrations in feeds have also been demonstrated to potentially lead to decreased DME selectivity and decreased CO_2 conversion. The dependence of CO conversion on the specific case is influenced by temperature [25]. CO conversion and DME output are anticipated to increase in slurry reactors, whereas direct DME synthesis is anticipated to result in decreased CO conversion in isothermal fixed-bed reactors. Additionally, to significantly increase the conversion of methanol to dimethyl ether, methanol can be dehydrated catalytically in an adiabatic fixed bed reactor; nevertheless, this approach diminishes the selectivity of the DME. It is determined that the synthesis of methanol is the limiting phase of the process, and the stoichiometry of the reaction indicates that increasing the pressure improves the CO conversion. Carbon monoxide (CO) conversion in both slurry and fixed-bed reactors was ultimately observed to be negatively impacted by space velocity. Despite the abundance of scholarly works dedicated to DME production, there is currently no established research domain that comprehensively addresses the economic aspects of the process.

Furthermore, it is critical to contemplate the optimization and expansion of the one-step synthesis of DME. The predominant function of CO_2 in the synthesis of DME has not been extensively studied, particularly in environments characterized by high velocity [59]. Furthermore, a consensus regarding the optimal CO/CO_2 ratio for the input stream has yet to be reached. Additional research is required to develop catalysts that exhibit enhanced activity, selectivity, and stability when exposed to water. To produce DME/MeOH mixtures that meet the specifications of various products, it is additionally essential to consider the optimal proportions of catalyst components. Additionally, it is recommended to confirm whether Lewis sites can transform into Bronsted sites when water is present. Additionally, it is supplied so that the optimal water removal value for membrane reactors can be determined. Additional recommendations for the implementation of this approach encompass conducting research on the efficacy of catalyst immobilization in microreactors and developing a robust catalyst specifically designed for the CH_3Br hydrolysis process.

References

[1] Staffell I, Scamman D, Velazquez Abad A, Balcombe P, Dodds PE, Ekins P, et al. The role of hydrogen and fuel cells in the global energy system. *Energy Environ Sci* 2019;12:463–491. https://doi.org/10.1039/C8EE01157E.

[2] Putrasari Y, Lim O. Dimethyl ether as the next generation fuel to control nitrogen oxides and particulate matter emissions from internal combustion engines: A review. *ACS Omega* 2022;7:32–37. https://doi.org/10.1021/acsomega.1c03885.

[3] Banivaheb S, Pitter S, Delgado KH, Rubin M, Sauer J, Dittmeyer R. Recent progress in direct DME synthesis and potential of bifunctional catalysts. *Chem Ing Tech* 2022;94:240–255. https://doi.org/10.1002/cite.202100167.

[4] Bateni H, Able C. Development of heterogeneous catalysts for dehydration of methanol to dimethyl ether: A review. *Catal Ind* 2019;11:7–33. https://doi.org/10.1134/S2070050419010045.

[5] Liu Y, Kirchberger FM, Müller S, Eder M, Tonigold M, Sanchez-Sanchez M, et al. Critical role of formaldehyde during methanol conversion to hydrocarbons. *Nat Commun* 2019;10. https://doi.org/10.1038/s41467-019-09449-7.

[6] Zhou Z, Liu H, Ni Y, Wen F, Chen Z, Zhu W, et al. Direct conversion of dimethyl ether and CO to acetone via coupling carbonylation and ketonization. *J Catal* 2021;396:360–373. https://doi.org/10.1016/j.jcat.2021.03.006.

[7] Kang J, He S, Zhou W, Shen Z, Li Y, Chen M, et al. Single-pass transformation of syngas into ethanol with high selectivity by triple tandem catalysis. *Nat Commun* 2020;11. https://doi.org/10.1038/s41467-020-14672-8.

[8] Serov A, Kwak C. Progress in development of direct dimethyl ether fuel cells. *Appl Catal B Environ* 2009;91:1–10. https://doi.org/10.1016/j.apcatb.2009.06.018.

[9] Mondal U, Yadav GD. Perspective of dimethyl ether as fuel: Part I. Catalysis. *J CO2 Util* 2019;32:299–320. https://doi.org/10.1016/j.jcou.2019.02.003.

[10] Azizi Z, Rezaeimanesh M, Tohidian T, Rahimpour MR. Dimethyl ether: A review of technologies and production challenges. *Chem Eng Process Process Intensif* 2014;82:150–172. https://doi.org/10.1016/j.cep.2014.06.007.

[11] Ua Cearnaigh DC, Dumont JH, Chung HT, Zelenay P. High-performance direct dimethyl ether fuel cell operating with an advanced activation process. *ECS Meet Abstr* 2015;MA2015–02:1365–1365. https://doi.org/10.1149/ma2015–02/37/1365.

[12] Herron JA, Ferrin P, Mavrikakis M. First-principles mechanistic analysis of dimethyl ether electro-oxidation on monometallic single-crystal surfaces. *J Phys Chem C* 2014;118:24199–24211. https://doi.org/10.1021/jp505919x.

[13] Nilsson M, Jansson K, Jozsa P, Pettersson LJ. Catalytic properties of Pd supported on ZnO/ZnA12O4/A12O3 mixtures in dimethyl ether autothermal reforming. *Appl Catal B Environ* 2009;86:18–26. https://doi.org/10.1016/j.apcatb.2008.07.012.

[14] Mizutani I, Liu Y, Mitsushima S, Ota K-ichiro, Kamiya N. Anode reaction mechanism and crossover in direct dimethyl ether fuel cell. *J Power Sources* 2006;156:183–189. https://doi.org/10.1016/j.jpowsour.2005.05.041.

[15] Abdel-Hady EE, Gamal A, Hamdy H, Shaban M, Abdel-Hamed MO, Mohammed MA, et al. Methanol electro oxidation on Ni–Pt–CrO/CNFs composite: Morphology, structural, and electrochemical characterization. *Sci Rep* 2023;13:1–16. https://doi.org/10.1038/s41598-023-31940-x.

[16] Beltermann N, Weiske S, Becka R, Samsun RC, Peters R, Stolten D, et al. An evaluation of direct dimethyl ether (DME) synthesis from hydrogen and carbon dioxide based on CFD reactor simulations. *Int J Hydrogen Energy* 2023;48:39373–39388. https://doi.org/10.1016/j.ijhydene.2023.05.260.

[17] Kéranguéven G, Coutanceau C, Sibert E, Hahn F, Léger JM, Lamy C. Mechanism of di(methyl)ether (DME) electrooxidation at platinum electrodes in acid medium. *J Appl Electrochem* 2006;36:441–448. https://doi.org/10.1007/s10800-005-9095-6.

[18] Liu Y, Muraoka M, Mitsushima S, Ota KI, Kamiya N. Electrochemical and ATR-FTIR study of dimethyl ether and methanol electro-oxidation on sputtered Pt electrode. *Electrochim Acta* 2007;52:5781–5788. https://doi.org/10.1016/j.electacta.2007.02.061.

[19] Zhang Y, Lu L, Tong Y, Osawa M, Ye S. Electrochemical and infrared study of electro-oxidation of dimethyl ether (DME) on platinum polycrystalline electrode in acid solutions. *Electrochim Acta* 2008;53:6093–6103. https://doi.org/10.1016/j.electacta.2008.01.109.
[20] Luu MT, Milani D, Wake M, Abbas A. Analysis of di-methyl ether production routes: Process performance evaluations at various syngas compositions. *Chem Eng Sci* 2016;149:143–155. https://doi.org/10.1016/j.ces.2016.04.019.
[21] Huang MH, Lee HM, Liang KC, Tzeng CC, Chen WH. An experimental study on single-step dimethyl ether (DME) synthesis from hydrogen and carbon monoxide under various catalysts. *Int J Hydrogen Energy* 2015;40:13583–13593. https://doi.org/10.1016/j.ijhydene.2015.07.168.
[22] Deka TJ, Osman AI, Baruah DC, Rooney DW. Methanol fuel production, utilization, and techno-economy: A review. *Environ Chem Lett* 2022;20:3525–3554. https://doi.org/10.1007/s10311-022-01485-y.
[23] Makos P, Słupek E, Sobczak J, Zabrocki D, Hupka J, Rogala A. Dimethyl ether (DME) as potential environmental friendly fuel. *E3S Web Conf* 2019;116. https://doi.org/10.1051/e3sconf/201911600048.
[24] Lotfollahzade Moghaddam A, Hazlett MJ. Methanol dehydration catalysts in direct and indirect dimethyl ether (DME) production and the beneficial role of DME in energy supply and environmental pollution. *J Environ Chem Eng* 2023;11. https://doi.org/10.1016/j.jece.2023.110307.
[25] Stiefel M, Ahmad R, Arnold U, Döring M. Direct synthesis of dimethyl ether from carbon-monoxide-rich synthesis gas: Influence of dehydration catalysts and operating conditions. *Fuel Process Technol* 2011;92:1466–1474. https://doi.org/10.1016/j.fuproc.2011.03.007.
[26] Dieterich V, Buttler A, Hanel A, Spliethoff H, Fendt S. Power-to-liquid via synthesis of methanol, DME or Fischer–Tropsch-fuels: A review. *Energy Environ Sci* 2020;13:3207–3252. https://doi.org/10.1039/d0ee01187h.
[27] Hussien AGS, Polychronopoulou K. A review on the different aspects and challenges of the dry reforming of methane (DRM) reaction. *Nanomaterials* 2022;12. https://doi.org/10.3390/nano12193400.
[28] Ardy A, Pohan RDA, Rizkiana J, Laniwati M, Susanto H. Dehydration of methanol to dimethyl ether (DME): Performance of three types of catalyst at atmospheric pressure. *AIP Conf Proc* 2019;2085. https://doi.org/10.1063/1.5095042.
[29] Palomo J, Rodríguez-Cano MA, Rodríguez-Mirasol J, Cordero T. On the kinetics of methanol dehydration to dimethyl ether on Zr-loaded P-containing mesoporous activated carbon catalyst. *Chem Eng J* 2019;378. https://doi.org/10.1016/j.cej.2019.122198.
[30] Hadi GJ, Hadi AJ. Kinetic study of methanol dehydration to dimethyl ether in catalytic packed bed reactor over resin. *J Mater Sci Chem Eng* 2022;10:45–58. https://doi.org/10.4236/msce.2022.106005.
[31] Yaripour F, Baghaei F, Schmidt I, Perregaard J. Catalytic dehydration of methanol to dimethyl ether (DME) over solid-acid catalysts. *Catal Commun* 2005;6:147–152. https://doi.org/10.1016/j.catcom.2004.11.012.
[32] Bandiera J, Naccache C. Kinetics of methanol dehydration on dealuminated H-mordenite: Model with acid and basic active centres. *Appl Catal* 1991;69:139–148. https://doi.org/10.1016/S0166–9834(00)83297–2.

[33] Wang Y, Wang WL, Chen YX, Zheng JJ, Li RF. Synthesis of dimethyl ether from syngas using a hierarchically porous composite zeolite as the methanol dehydration catalyst. *Ranliao Huaxue Xuebao/J Fuel Chem Technol* 2013;41:875–882. https://doi.org/10.1016/s1872–5813(13)60037–7.

[34] You Q, Liu Z, Li W, Zhou X. Synthesis of dimethyl ether from methane mediated by HBr. *J Nat Gas Chem* 2009;18:306–311. https://doi.org/10.1016/S1003–9953(08)60122-X.

[35] Ereña J, Sierra I, Aguayo AT, Ateka A, Olazar M, Bilbao J. Kinetic modelling of dimethyl ether synthesis from (H2+CO2) by considering catalyst deactivation. *Chem Eng J* 2011;174:660–667. https://doi.org/10.1016/j.cej.2011.09.067.

[36] Iliuta I, Iliuta MC, Larachi F. Sorption-enhanced dimethyl ether synthesis-Multiscale reactor modeling. *Chem Eng Sci* 2011;66:2241–2251. https://doi.org/10.1016/j.ces.2011.02.047.

[37] Prakash GKS, Colmenares JC, Batamack PT, Mathew T, Olah GA. Poly(4-vinylpyridine) catalyzed hydrolysis of methyl bromide to methanol and dimethyl ether. *J Mol Catal A Chem* 2009;310:180–183. https://doi.org/10.1016/j.molcata.2009.06.018.

[38] Zha F, Tian H, Yan J, Chang Y. Multi-walled carbon nanotubes as catalyst promoter for dimethyl ether synthesis from CO 2 hydrogenation. *Appl Surf Sci* 2013;285:945–951. https://doi.org/10.1016/j.apsusc.2013.06.150.

[39] Zhang Q, Zuo YZ, Han MH, Wang JF, Jin Y, Wei F. Long carbon nanotubes intercrossed Cu/Zn/Al/Zr catalyst for CO/CO2 hydrogenation to methanol/dimethyl ether. *Catal Today* 2010;150:55–60. https://doi.org/10.1016/j.cattod.2009.05.018.

[40] Li Q, Wu G, Johnston CM, Zelenay P. Anode catalysts for the direct dimethyl ether fuel cell. *ECS Meet Abstr* 2011;MA2011–02:833–833. https://doi.org/10.1149/ma2011–02/16/833.

[41] Cai K Di, Yin GP, Wang JJ, Lu LL. Investigation of an anode catalyst for a direct dimethyl ether fuel cell. *Energy and Fuels* 2009;23:903–907. https://doi.org/10.1021/ef8005965.

[42] de Sá MH, Moreira CS, Pinto AMFR, Oliveira VB. Recent advances in the development of nanocatalysts for direct methanol fuel cells. *Energies* 2022;15:6335. https://doi.org/10.3390/EN15176335.

[43] Ramli ZAC, Kamarudin SK. Platinum-based catalysts on various carbon supports and conducting polymers for direct methanol fuel cell applications: A review. *Nanoscale Res Lett* 2018;13:1–25. https://doi.org/10.1186/S11671-018-2799-4.

[44] Osman SH, Kamarudin SK, Basri S, Karim NA. Organic aerogel as electro-catalytic support in low-temperature fuel cell. *Int J Energy Res* 2022;46:16264–16280. https://doi.org/10.1002/ER.8377.

[45] Liu Y, Mitsushima S, Ota K ichiro, Kamiya N. Electro-oxidation of dimethyl ether on Pt/C and PtMe/C catalysts in sulfuric acid. *Electrochim Acta* 2006;51:6503–6509. https://doi.org/10.1016/j.electacta.2006.04.037.

[46] Kerangueven G, Coutanceau C, Sibert E, Léger JM, Lamy C. *Methoxy methane (dimethyl ether) as an alternative fuel for direct fuel cells. J Power Sources* 2006;157:318–324. https://doi.org/10.1016/j.jpowsour.2005.07.080.

[47] Mench MM, Chance HM, Wang CY. Direct dimethyl ether polymer electrolyte fuel cells for portable applications. *J Electrochem Soc* 2004;151:A144. https://doi.org/10.1149/1.1631819.

[48] Cai K Di, Yin GP, Zhang J, Wang ZB, Du CY, Gao YZ. Investigation of a novel MEA for direct dimethyl ether fuel cell. *Electrochem Commun* 2008;10:238–241. https://doi.org/10.1016/j.elecom.2007.11.035.

[49] Bakhtyari A, Parhoudeh M, Rahimpour MR. Optimal conditions in converting methanol to dimethyl ether, methyl formate, and hydrogen utilizing a double membrane heat exchanger reactor. *J Nat Gas Sci Eng* 2016;28:31–45. https://doi.org/10.1016/j.jngse.2015.11.028.

[50] Kumar M, Srivastava VC. Simulation of a fluidized-bed reactor for dimethyl ether synthesis. *Chem Eng Technol* 2010;33:1967–1978. https://doi.org/10.1002/ceat.201000158.

[51] Lu WZ, Teng LH, Xiao W De. Simulation and experiment study of dimethyl ether synthesis from syngas in a fluidized-bed reactor. *Chem. Eng. Sci* 2004;59:5455–5464. https://doi.org/10.1016/j.ces.2004.07.031.

[52] Ereña J, Garoña R, Arandes JM, Aguayo AT, Bilbao J. Direct synthesis of dimethyl ether from (H2+CO) and (H 2+CO2) feeds. Effect of feed composition. *Int J Chem React Eng* 2005;3. https://doi.org/10.2202/1542–6580.1295.

[53] Ereña J, Garoña R, Arandes JM, Aguayo AT, Bilbao J. Effect of operating conditions on the synthesis of dimethyl ether over a CuO-ZnO-Al2O3/NaHZSM-5 bifunctional catalyst. *Catal. Today* 2005;107–108:467–473. https://doi.org/10.1016/j.cattod.2005.07.116.

[54] Ateka A, Pérez-Uriarte P, Sánchez-Contador M, Ereña J, Aguayo AT, Bilbao J. Direct synthesis of dimethyl ether from syngas on CuO[sbnd]ZnO[sbnd]MnO/SAPO-18 bifunctional catalyst. *Int J Hydrogen Energy* 2016;41:18015–18026. https://doi.org/10.1016/j.ijhydene.2016.07.195.

[55] Khoshbin R, Haghighi M. Direct syngas to DME as a clean fuel: The beneficial use of ultrasound for the preparation of CuO-ZnO-Al2O3/HZSM-5 nanocatalyst. *Chem Eng Res Des* 2013;91:1111–1122. https://doi.org/10.1016/j.cherd.2012.11.017.

[56] Zhou X, Su T, Jiang Y, Qin Z, Ji H, Guo Z. CuO-Fe2O3-CeO2/HZSM-5 bifunctional catalyst hydrogenated CO2 for enhanced dimethyl ether synthesis. *Chem Eng Sci* 2016;153:10–20. https://doi.org/10.1016/j.ces.2016.07.007.

[57] Liu R, Zhu F, Steinberger Y. Effectiveness of afforested shrub plantation on ground-active arthropod communities and trophic structure in desertified regions. *Catena* 2015;125:1–9. https://doi.org/10.1016/j.catena.2014.09.018.

[58] Lee S, Gogate MR, Kulik CJ. A novel single-step dimethyl ether (DME) synthesis in a three-phase slurry reactor from co-rich syngas. *Chem Eng Sci* 1992;47:3769–3776. https://doi.org/10.1016/0009–2509(92)85096-T.

[59] Bonura G, Cannilla C, Frusteri L, Mezzapica A, Frusteri F. DME production by CO2 hydrogenation: Key factors affecting the behaviour of CuZnZr/ferrierite catalysts. *Catal Today* 2017;281:337–344. https://doi.org/10.1016/j.cattod.2016.05.057.

7

System Design and Commercialization Aspect of Direct Liquid Fuel Cell Technology

Mohd Saqib, Zulnurain Khan, Amjad Mumtaz Khan, Ansari M.N.M., and Norsyahida Mohammad

7.1 Introduction

The increasing awareness of environmental sustainability has driven promising alternatives to fossil fuels, thrusting new renewable and clean energy alternatives, such as hydro, solar, and wind energy [1,2]. The intermittency nature of these new energy technologies powered the research and development of other promising power generation technologies, including fuel cells, which were first introduced in the 1950s due to the development of space exploration programs [3–5]. In the 1970s, fuel cell developments focused on large solutions particularly for power generation, partly due to the oil crisis [6,7]. Since the 1990s, the focus shifted to specific applications, including portable power solutions, creating a range of fuel cell technology, such as hydrogen fuel cells, also known as proton exchange membrane fuel cell or polymer electrolyte membrane fuel cell (PEMFC), solid oxide fuel cell (SOFC), molten carbonate fuel cell (MCFC), alkaline fuel cell (AFC), and phosphoric acid fuel cell (PAFC) [5,8–10]. DLFC is a sub-category of PEMFC. DLFCs have been one of the most prominent fuel cells for portable applications due to their ability to produce medium to high power range (100 W to 3000 W), ease of liquid fuel storage, and ease of refuelling compared to other types of fuel cells [11,12]. DLFCs could also be utilized for power generation, provided the abundance of fuel. However, the power density and energy density of DLFC is not sufficient for utilization in automobile industries [11,13]. Therefore, a reformat methanol fuel cell (RMFC) was introduced, where methanol is reformed to produce hydrogen-rich gas prior feeding into the PEMFC [6,14,15].

In 2022, the global market of DLFCs is dominated by four major companies—SFC Energy AG (Germany), Oorja Corporation (United States), Fujikura Ltd. (Japan), and SIQENS GmbH (Germany)—mainly for mobility

DOI: 10.1201/9781003512592-7

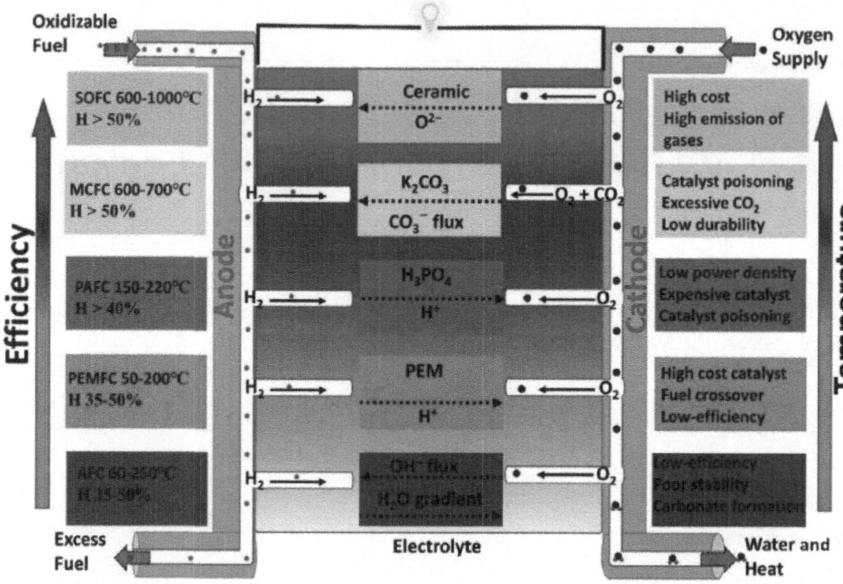

FIGURE 7.1
Schematic illustration of different types of fuel cells operating at different temperatures, efficiency, and their challenging issues [19].

and leisure applications, while the military and defense sectors require a more stringent standard. The commercialization of DLFCs is influenced by its system design, which includes regulated fuel supply, water management, and heat dissipation to maintain high performance [11,13,16]. The fuel cell should also be compact and lightweight while maintaining its cost competitiveness to facilitate commercialization [5,17,18]. Figure 7.1 shows how several fuel cell designs are compared according to factors, including cost, temperature, and current challenges. PEM is further classified as "inorganic PEMs," "organic PEMs," and "composite PEMs" in the Scopus database. Figure 7.2 shows the total number of papers pertaining to PEM from 2017 to 2023.

7.2 Advancement in System Design of DLFC

Although DLFC remains the most "simple fuel cell system," there arise general issues related to the system design, including (i) sluggish anode

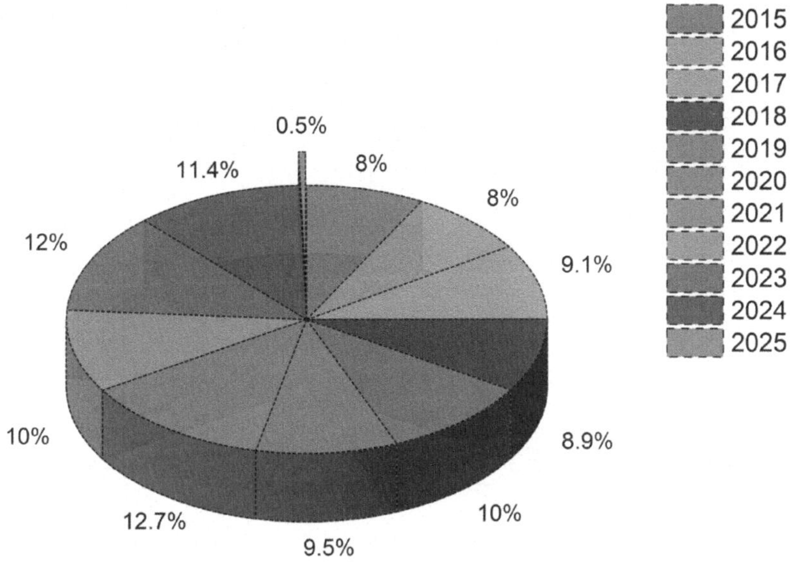

FIGURE 7.2
Total number of publications per year (accessed from the Scopus database 2015–2025).

kinetics due to the multiple-step fuel oxidation at the anode, (ii) fuel crossover through the polymer electrolyte membrane causing power loss, degradation of cathode performance, and generation of extra heat, (iii) anode gaseous product removal, and (iv) water recovery. The main fuel of DLFC methanol contains anode which produces the byproduct of liquid fuel oxidation because of incomplete oxidation, whereas the cathode waste product turns into water (Figure 7.3) and is often associated with its toxicity, requiring extra safety precautions [20]. Alternatively, ethanol emerges as a promising fuel alternative due to its non-toxicity and abundance, albeit lower power density [21,22]. In addition, other liquid fuels, such as formic acid, can generate higher power densities, although they are more expensive than methanol [23,24]. Even though various initiatives were taken to explore fuel alternatives for DLFCs, the system design of DLFCs is always a fundamental technical challenge in producing high-performance, cost-competitive, and compact-sized DLFCs, which are often referred to as micro or small DLFCs [11,20].

The overall system design of a DLFC includes the fuel supply mode, the basic unit cell of the fuel cell stack, the flow field design, and the membrane electrode assembly (MEA) design [11,25,26]. Include explanation of BOP similar to Faghri 2008 ("A DMFC system includes a fuel cell stack, a fuel tank and

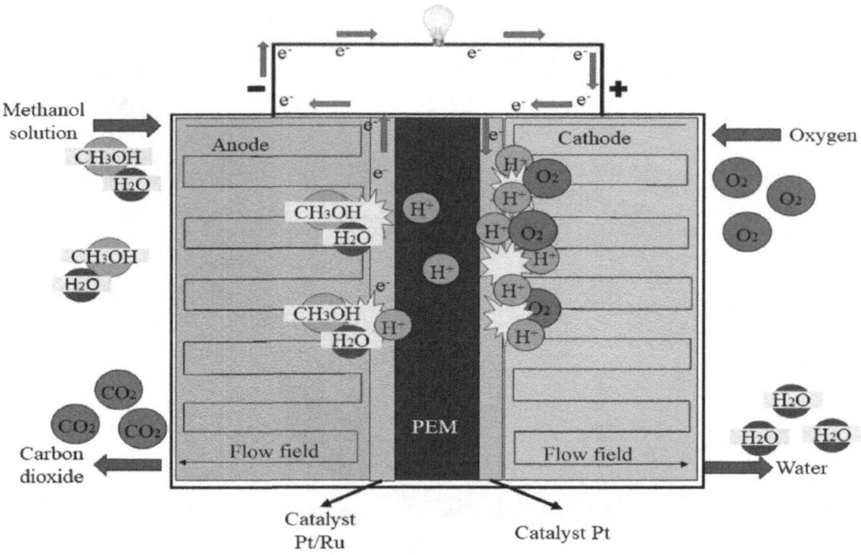

FIGURE 7.3
Reactions occurring in the DMFC fuel cell system.

Source: Reproduced with permission from [11].

balance of plant (BOP). The BOP includes all the additional parts of a fuel cell system that support the power generation process. Some common BOP components are pumps, fans, valves, sensors, control circuits, plus many more, depending on the system.") and overall drawing of BOP. Tremendous efforts were made to enhance the performance of each of its components, particularly in improving the performance of the MEAs, while maintaining lower costs [27,28].

7.2.1 Fuel Supply Mode

Earlier research and developments of fuel cells discovered the high performance of fuel cells fed by vaporized fuel. However, due to its complexity, recent studies have focused more on liquid-feed DLFCs [29,30]. The two main modes for fuel supply in DLFCs are the passive and active modes. Passive fuel supply relies on capillary forces, diffusion, air convection, or evaporation to deliver fuel to the reaction sites without external pumps [31,32]. This method is advantageous for portable applications due to its simplicity, without the requirement of additional power consumption [33]. In contrast, active fuel supply methods use additional pieces of equipment, including a pump and fan, ensuring a controlled flow and consistent concentration of fuel to achieve desired operating conditions, such as temperature and pressure,

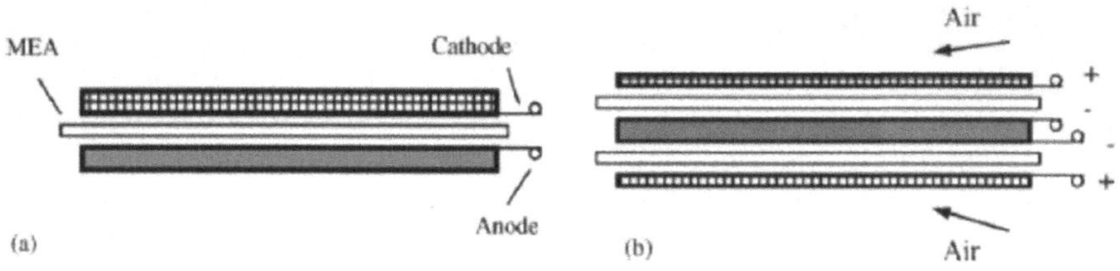

FIGURE 7.4
Basic unit cell configurations: (a) unit single cell; (b) unit bi-cell [39].

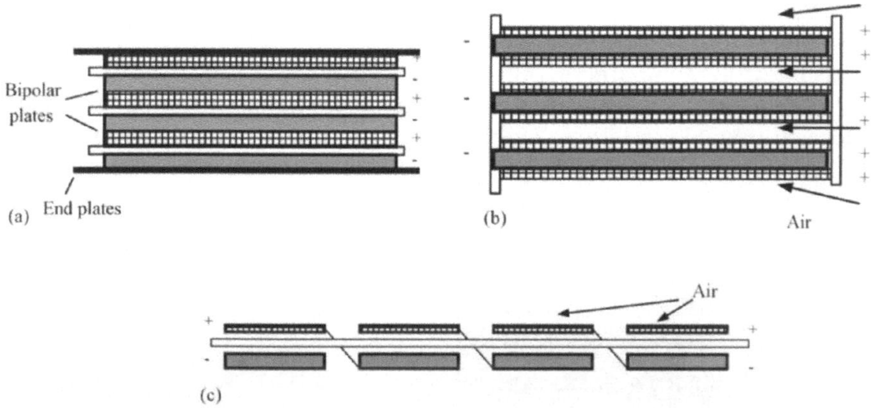

FIGURE 7.5
Stack configurations: (a) bipolar parallel; (b) bi-cell parallel; (c) mono-polar strip [39].

which can improve overall fuel cell performance. The latter is more suited to larger fuel cells and stationary applications [34,35].

7.2.2 Basic Unit and Stack Configuration

The design of a basic unit and stack configuration are critical for the overall performance and scalability of DLFCs [20,36]. A basic unit cell can be designed as a single cell or a bi-cell, depicted in Figure 7.4 and 7.5. A bi-cell configuration provides access to air at the outside-facing cathodes. These two configurations may be combined to produce three stack configurations: bipolar stack, bi-cell stack, and mono-polar strip stack, as presented in Figure 7.3. Different cell designs include planar and tubular configurations, which can be assembled into stacks to achieve the desired power output [20,36]. The planar design is common due to its simplicity and ease of manufacturing. Stacks can also be configured in series or parallel arrangements to optimize voltage and current output, addressing specific application requirements [37–39].

7.2.3 Flow Field Design

Flow field design plays a crucial role in the efficient distribution of reactants and removal of byproducts in DLFCs [11,40]. Effective anode gas removal is crucial to maintain continuous fuel supply and prevent the accumulation of gases that can cause pressure build-up and reduce cell efficiency [41,42]. An optimal design of the anode flow field and the incorporation of suitable gas diffusion layers can help in efficient gas removal, ensuring steady-state operation of the fuel cell [43,44]. There are various flow field patterns, including serpentine, parallel, and interdigitated designs. These patterns influence the performance by affecting the pressure drop, reactant distribution, and water management within the cell [45,46]. The serpentine flow field is effective in ensuring uniform distribution of fuel and oxidant but may result in higher pressure drops compared to other designs [43,47,48].

In this work, two types of 3D flow fields were designed for PEMFC cathode, as illustrated in Figure 7.6, both of which are three-channel serpentine

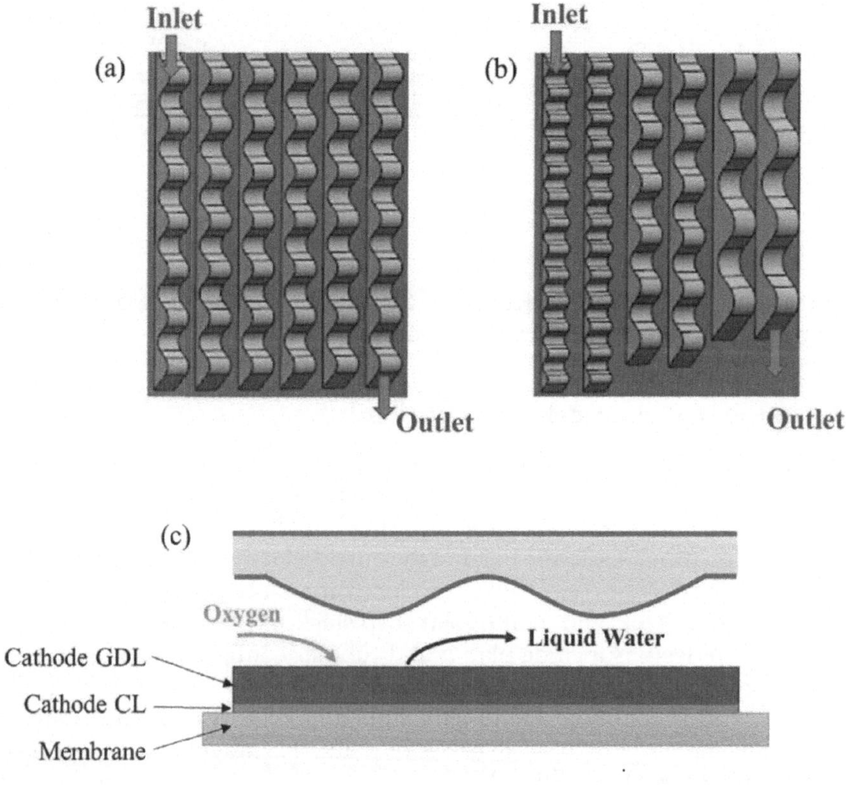

FIGURE 7.6
Schematic of flow fields with different 3D geometry [43].

flow patterns. The first one (3D flow field, Figure 7.6a) is designed to own waved channels to induce a through-plane convection of oxidant. The other one (gradient 3D flow field, Figure 7.6b) also has 3D waved channels, with the channel depth gradually decreasing from the upstream region to the downstream region. This design increases flow velocity at both the in-plane and through-plane directions, helping to overcome severe water flooding in the downstream region. Conventional three-channel serpentine flow field is used as a control group, which is denoted as 2D flow field hereinafter [43].

7.2.4 Membrane Electrode Assembly (MEA) Design

The membrane electrode assembly (MEA) is the core component of DLFCs, where electrochemical reactions occur, converting chemical energy from the fuel into electrical power. The MEA is a proton exchange membrane (PEM) sandwiched between an anode and a cathode, each coated with catalysts [49–51]. A zero-gap cell, or MEA-configuration cell, is an important component in lowering the system's ohmic loss. Essential components, such as the anode plate, anode gas diffusion layer (GDL), anode catalytic layer (CL), cathode gas diffusion layer, cathode plate, and polymer electrolyte membranes (PEM), are shown in Figure 7.7a. An electrochemical stack device may be expanded

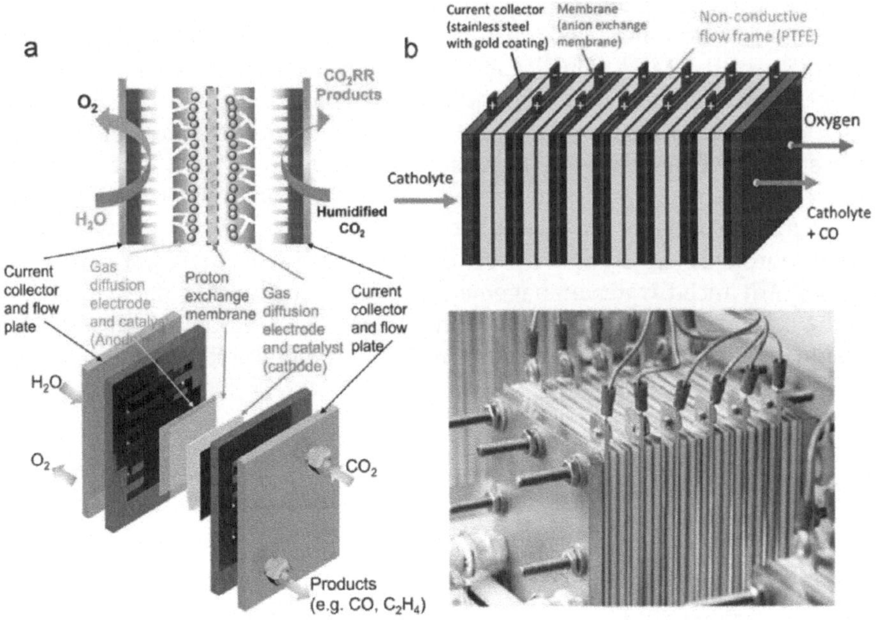

FIGURE 7.7
(a) Schematic of MEA cell (copyright © 2022, Elsevier) and (b) MEA stack for CO_2 electrocatalysis (copyright © 2015, Delaware University and NASA Glenn Research Centre) [51].

into many configurations with the help of MEA. As shown in Figure 7.7b, a six-cell stack was created by Jiao et al. This stack consists of six distinct cells that are electrically and fluidically coupled in parallel and are spaced apart by Teflon sheets [51].

The most common electro-catalysts for DLFC applications, particularly in direct methanol fuel cells (DMFCs), include platinum (Pt) and platinum alloyed with transition metals, such as ruthenium (PtRu), with/without the support material, such as graphene or carbon [52,53]. The choice of materials and the optimization of the catalyst layer are vital to prevent sluggish anode kinetics, improving performance and reducing costs [54,55]. These issues can be addressed by developing more active catalysts and optimizing the catalyst layer structure. The alloyed catalyst is a better option because it can withstand carbon monoxide (CO) poisoning and improves both the stability and the activity of the fuel cell, reducing overall cost [56,57].

Another issue related to the MEA is the fuel crossover, particularly methanol crossover in DMFCs. Methanol crossover from the anode to the cathode reduces fuel efficiency and can lead to mixed potentials at the cathode [58–60]. Strategies to mitigate this include optimizing membrane thickness, utilizing hybrid membranes, applying membrane surface modifications or coatings to enhance methanol-blocking properties, and innovating the MEA so that barrier layers could be incorporated to help reduce methanol crossover while maintaining proton conductivity [58–60].

Water is required for the anode reaction but is also produced at the cathode. Cathode flooding needs to be addressed to enhance the efficiency and longevity of the MEA [61,62]. Cathode flooding occurs when excess liquid water accumulates in the cathode catalyst layer and gas diffusion layer (GDL), leading to blockage of the pores and hindering the flow of reactant gases [63,64]. This results in decreased performance due to limited oxygen access and increased mass transport losses [65]. Proper water management is essential to mitigate this issue, involving the use of hydrophobic GDLs and micro-porous layers to control water distribution and facilitate water removal [66,67]. Advanced water management involves reusing the water produced at the cathode for the anode reaction, which can be achieved through the design of a water management system within the fuel cell [68,69].

7.3 Fuel Cell Commercialization

The milestone towards DLFC commercialization can be seen by the efforts made in research and development, which have materialized in substantial journal publications and patent filings from the early 1990s. Tokyo's Toshiba Corporation filed the highest number of patent applications, and in 2004, they announced a compact DMFC prototype that can be incorporated into small devices, such as digital audio players and mobile wireless headsets [70,71]. In 2009, Toshiba's first commercial DMFC mobile charger was launched [72]. The portable DLFC units shipped between 2005 and 2008 quadrupled, with China and Taiwan focusing mainly on technological demonstrations, while the European countries and the United States companies ventured into military applications. However, there are several barriers to be addressed before further commercialization. These barriers are cost-competitiveness and performance, manufacturing processes, policy and standards, consumer habits, and application trends [11,73].

An electrochemical reaction is a process that changes the energy of chemicals into electricity. Fuel cells are a more modern and efficient way to generate power than traditional combustion engines. Depending on the specific kind of fuel cell, its potential uses might differ. Gaining a comprehensive understanding of the various fuel cell types is essential for fully using their capabilities in diverse businesses and sectors. We will provide a thorough introduction to this state-of-the-art technology by reviewing the features and uses of several fuel cell types in this review article. The element budgets for a fuel cell stack produced at a yearly rate of 500,000 units are shown in Figure 7.8. Research shows that the catalyst layer is the most expensive part, making about 41% of the total [74].

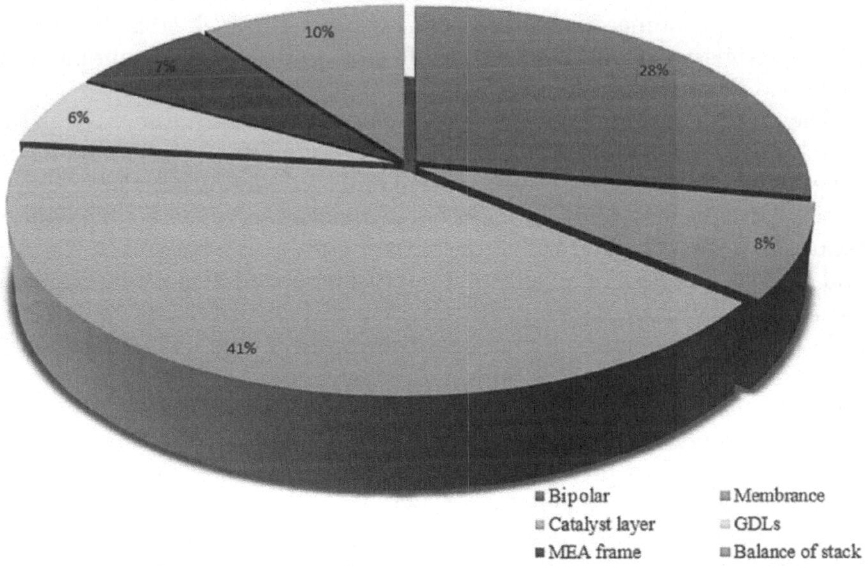

FIGURE 7.8
Breakdown of component costs for an annual fuel cell stack production run of 500,000 units [74].

Scholars greatly aspire the use of DLFC as an alternative system that can work better than lithium-ion batteries and replace conventional fossil fuel–based systems. Lithium-ion batteries have low energy density and can only operate within a short time. Meanwhile, fossil fuel–based systems generate high energy density but may cause environmental pollution. In this regard, the importance and potential of DLFCs are undeniable and have been the focus in the industry and research. DLFCs represent a promising energy technology that has gained attention for its potential in applications ranging from portable electronics to electric vehicles and backup power systems. Their ability to directly convert liquid fuels, like methanol, ethanol, or formic acid, into electricity without intermediate reforming steps provides several advantages, including higher efficiency, reduced system complexity, and portability. Despite these benefits, the commercialization of DLFCs is still in its early stages due to challenges related to cost, performance, and durability [75]. A passive DMFC costs more than a Li-ion battery due to the usage of precious platinum catalyst at both electrodes to attain higher power density. Since noble metal-based catalysts in DMFCs contribute to approximately 40% of its cost, non-noble metal catalyst alternatives and reduced noble metal loading levels are practical approaches to assist the commercialization of DMFCs. However, non-noble metal catalysts cannot compete with noble platinum catalysts due to their lower activities. Platinum (Pt) and Au have also been utilized as the current collector coating to enhance corrosion resistance and to reduce contact resistance between the MEA and current collector. Costly Nafion membrane in DMFCs also contributes to the overall cost. Although numerous studies have been done to seek suitable alternative materials for catalysts, current collector coating, and membrane, cost remained the major hurdle for DLFC commercialization [20].

The fuel crossover phenomenon, where liquid fuels permeate through the membrane, reduces efficiency and durability. As fuel crossover occurs for all types of DLFCs and reduces cell performance, some innovation is necessary to solve this serious problem. Dilution of the fuel seems to be one way to reduce the crossover rate, as the concentration difference between the anode and cathode has been reduced, but more fuel is needed to generate the same amount of power as concentrated fuel. The fuel crossover rate can also be reduced using a membrane that is less permeable to the fuel. Maabet et al. used a modified sulphonated poly(ether ether ketone), or SPEEK, membrane for DEFCs, and a lower crossover rate of ethanol was achieved [11]. There is still much effort needed to totally eliminate the fuel crossover problem in DLFCs to increase the utilization of the fuel and improve the overall cell performance [11,76]. Scalability is another consideration for DLFC commercialization. To penetrate mass markets, manufacturers need to transition from small-scale production to high-volume manufacturing. Initiatives to establish robust supply chains, automate production processes, and integrate standardized components are crucial to achieving this goal. Furthermore, the establishment of infrastructure for the distribution and storage of liquid fuels is essential to support widespread adoption. Portable electronics, unmanned aerial vehicles (UAVs), and backup power systems are ideal initial applications due to their smaller scale and premium performance requirements. As these markets mature, DLFCs can potentially expand into automotive and stationary applications, provided costs decrease and performance improves. Strategic partnerships, government incentives, and funding for pilot projects play pivotal roles in accelerating commercialization efforts [25,26].

7.3.1 Cost-Competitiveness and Performance

DLFC is mostly compared to Li-ion batteries in assessing its potential for commercialization. A study suggested that by neglecting the weight and volume of DLFC, the performance of a DLFC is comparable to a Li-ion battery in providing power for a laptop computer with a 20 W power output and a 3000 h operational lifetime [20]. However, the specific energy and energy density of DLFCs based on the total system weight and volume were less competitive than Li-ion batteries. The high cost of the catalyst and high catalyst loading prevent DLFCs from being price competitive when compared to state-of-art power supplies, such as lithium-ion batteries. Until the current research effort, most DLFCs still required the application of an expensive noble catalyst to facilitate the cell reaction, and their loading was high compared to that of hydrogen-fed PEMFCs. For example, in the case of DMFCs, PtRu was denoted as the best catalyst for the anode, and Pt was the best for the cathode. A high peak power density of 23 mW cm^{-2} was obtained by using 8 mg cm^{-2} of catalyst on both of the electrodes for passive DMFCs [77]. Work on the reduction of catalyst loading had been performed by several researchers. For example, Ahmad et al. used 5 mg cm^{-2} of PtRu as the anode catalyst for DMFCs and achieved 14 mW cm^{-2}, while Hashim et al. found that by further reducing the catalyst loading to 3 mW cm^{-2}, the peak power density for DMFCs obtained was only 12 mW cm^{-2} [11,78]. Gogel et al. reported that reduction in the catalyst loading lead to a concurrent reduction in performance [79]. Although many different types of catalysts besides PtRu have been proposed for DMFCs, Pt still remains as one of the bimetallic or trimetallic elements. Pereria et al. used 4 mg cm^{-2} of PtRu as the anode catalyst and 2 mg cm^{-2} Pt as the cathode catalyst for passive DEFCs and reported that a peak power density of 1.33 mW cm^{-2} was obtained [80]. Matos et al. used a high loading of Pt catalyst of 4 mg cm^{-2} to achieve great performance for DFAFCs [81]. High loading of catalyst means higher cost for the fuel cell, especially when using a noble metal catalyst, such as Pt. The current main challenge for DLFCs is to find a suitable solution to reduce catalyst loading or use an inexpensive non-noble catalyst, thereby reducing the fabrication costs [11,82].

One of the promising approaches for reducing the amount of catalyst loading is to improve the utilization of the catalyst. This improvement may be achieved by developing advanced supporting materials, such as carbon nanofibers, carbon nanotubes, mesoporous carbon, and graphene to increase the catalyst utilization [83]. One of the examples of such an effort was the work of Zainoodin et al., who compared the effects of different types of supporting materials onto the performance of DMFCs [84]. With the same catalyst loading of 8 mg cm^{-2} for both the anode and the cathode, a carbon-nanofiber-supported anode was able to provide a peak power density of 18.3 mW cm^{-2}, which was higher than that from a conventional carbon-black-supported anode of 12.6 mW cm^{-2} [85]. This result showed that carbon nanofibers can increase the catalyst utilization of DMFCs. Therefore, by finding a suitable catalyst-supporting material, the loading of catalyst can be reduced to achieve the same performance as that of the conventional supporting material. In addition, the catalyst loading may also be reduced by optimizing the electrode structure. Almheiri et al. presented a higher power density output of 104 mW cm^{-2} from the active system of DMFCs with a catalyst loading of PtRu 4 mg cm^{-2} [11]. However, Liu et al. reported that an ultrasonic spray method for the MEA fabrication process will change the pore structure of the catalyst [86]. The performance had been increased for the same amount of catalyst loading by alternating the fabrication method. Although some effort had been used to reduce the catalyst loading for DLFCs, currently, the loading was still high when compared to hydrogen-fed PEMFCs. Therefore, more research studies are required to solve the high catalyst loading issue and reduce the high cost of the noble catalyst used in DLFCs [11,13].

7.3.2 Manufacturing Processes

The manufacturing process of DLFCs is a multi-step operation that requires precision engineering to ensure high performance and durability. It begins with the production of the catalyst materials, often involving platinum or palladium, which are finely dispersed on carbon supports to maximize active surface area. The next step involves the fabrication of the membrane electrode assembly (MEA), which is the core component of a DLFC. Once the MEA is prepared, it is integrated into single cells along with flow field plates, gaskets, and other components that ensure proper fuel distribution and electrical contact. The assembly of multiple cells into a fuel cell stack is a critical step in scaling up the power output. Cells are connected in series or parallel, depending on the application requirements, and assembled into compact modules. This step involves meticulous alignment to ensure minimal resistive losses and uniform performance across the stack. Automated assembly lines are increasingly being employed to improve efficiency and reduce labor costs [8,10].

In the initial step towards the commercialization of DLFC, Toshiba has successfully developed the first DMFC prototype for use in computer laptops. The prototype capacity is 12 W of power generated, the operation time is five hours, and the system weighs only less than 1 kg [71]. LG Chem Ltd. increased the DMFC's potable capacity to up to 4,000 hours with 25 W; the fuel cartridge size is 200 mL, which can be used in laptops. Other companies that lead the use of portable DLFCs for laptops are NEC, Fujitsu, Panasonic, and Samsung Electronics [11]. DoCoMo developed a mobile phone charger, with a capacity of 5.4 V at 700 Ma, by using 30% methanol. The phone charger was improved through the collaboration between Fujitsu and DoCoMo by using a higher methanol concentration of 90%; the same energy was produced, but the volume of methanol was smaller [87]. Portable DLFC (Jenny 600), which was manufactured by Smart Fuel Cell Company, has also been applied in military equipment, especially for inland settlements. This portable DLFC facilitates military operations because it is 80% lighter than conventional power systems; moreover, its cartridge size is 350 mL and its power yield efficiency is 350 Wh^{-1}. Jenny 600 was upgraded to Jenny 1200 in 2013 by Smart Fuel Cell; the improved version has two times higher efficiency in power generation, has a smaller and lighter size, and more convenient to handle [11].

The cost of a custom-made fuel cell system ranges from $3,300 to $5,500 per kW. To be competitive, manufacturing costs need to be limited to $55 per kW for private cars, $220 to $330 per kW for commercial trucks, and $440 to $660 per kW for portable applications or power generation. While mass production may help lower costs, a significant portion of expenses related to raw materials, particularly noble catalysts like platinum and commercial membranes like Nafion, would not see substantial reductions through mass production. Another challenge before the mass production of DMFC is standardization requirements. Unlike batteries, DFMCs do not have a well-defined size; hence, standard dimensions or sizes well-suited to respective applications are to be determined before commercialization [41,47].

The United States Department of Energy (DOE) established specific standards for developing competitive DMFC components, including membranes with methanol permeability below 1×10^{-8} cm^2/s by 2015. However, further enhancements are necessary to achieve the desired performance and characteristics, including a (i) 5,000-hour membrane lifetime in passive DMFCs; (ii) low-platinum anode catalysts loading or high-performance non-platinum alternatives loading of less than 0.2 mg/cm^2; (iii) high-performance non-platinum cathode catalysts with a minimal metal loading of 0.2–0.5 mg/cm^2; and (iv) more oxidation-resistant non-carbon cathode supports. Efforts to further optimize the manufacturing process are ongoing, with emerging technologies such as 3D printing and advanced coating techniques enabling customization and rapid prototyping. These innovations, combined with economies of scale, are expected to lower production costs and make DLFCs more accessible for commercial applications. As manufacturing processes continue to evolve, they are likely to play a key role in driving the adoption of DLFCs in the global energy market [11,88].

7.3.3 Infrastructure and Policy Support

As a sub-category of PEMFC, the development and commercialization of DLFC is very much similar to a PEMFC or hydrogen fuel cell. PEMFCs fuelled by hydrogen require a more extensive hydrogen infrastructure, with significant challenges in hydrogen production, storage, and distribution due to high costs and safety concerns. Conversely, for DLFC, in particular, DMFC, the infrastructure for methanol production and distribution is relatively well-established, as methanol has been used for blending with gasoline due to its environmental and economic benefits [18,27]. While there are safety concerns regarding methanol's toxicity, a report by the Methanol Institute indicates that refuelling a methanol fuel cell car results in safe low-dose exposures (23–38 ppm for a few minutes), with a small intake of 3 milligrams of methanol. The primary concern with alcohols, including methanol, is toxicity from ingestion, skin or eye contact, or inhalation, where toxic intermediary products formaldehyde and formic acid are formed causing adverse health impacts to humans. However, similar to other fuels, including gasoline, accidental ingestion of methanol can be mitigated by appropriately designed fuel-dispensing systems [4,89].

The commercialization of DMFC is supported via the policies, incentives, and subsidies aimed to promote fuel cell, as observed in advanced countries, including Japan, the United States, South Korea, the European Union, and China. A vast majority of policies focused on developing PEMFCs, particularly hydrogen fuel cells for mobility applications. China, the largest automobile market in the world, announced its Alternative Energy Vehicles policy to mitigate severe smog and accelerate its economic growth. The policy covers various types of vehicles, with the main focus on electric vehicles, including fuel cell electric vehicles [5,9].

On another note, DMFC play a larger role in non-automotive applications, including material-handling equipment and stationary or portable power generation. Except the smallest applications for computers and other small electronic devices (generally less than 100 W in size), fuel cells are still reliant on government policies to compete successfully against established technologies. In the United States, two federal policies leading the market for fuel cells include the Investment Tax Credit of $3,000 per kW up to the maximum of 30% of the capital cost and the American Recovery and Reinvestment Act 2009, which allow fuel cell purchasers to obtain a subsidy in lieu of the tax credit. Several states offer additional incentives, including California, which introduced the Self-Generation Incentive Program. Other countries, such as Japan, which focused on fuel cell applications in combined heat and power (CHP), introduced government subsidies for utilities to purchase fuel cell CHP systems [20,90].

7.3.4 Consumer Habits and Application Trends

Multifunctional portable electronic devices requiring power for longer time period will be utilizing a DMFC charging system due to its high energy density, simplified refuelling, and less time consumption. Future requirements will be accomplished by designing a simplified system architecture, operating under ambient temperature and pressure associated with fewer peripheral elements. The industry and research have focused on DLFC because of its undeniable significance and potential. DMFC is ranked top in terms of maturity and industry recognition based on the ranking of each type of DLFC system, followed by DEFC [11,41]. Toshiba created the initial compact form factor DMFC prototype for a laptop computer. It can run continuously for five hours and has an average power output of 12 W with a fuel cartridge weighing less than 1 kg [9,91]. In order to power a 25 W laptop for more than ten hours using a 200 ml fuel cartridge, LG Chem Ltd. developed portable DMFCs for laptop computers that have a lifespan of more than 4,000 hours. Leading firms that developed DMFCs as laptop computer power sources were NEC, Fujitsu, Panasonic, and Samsung Electronics. DoCoMo created DMFCs that produced 5.4 V at 700 mA for a mobile phone charger that contained 30% methanol. DoCoMo and Fujitsu worked together to enhance the DMFCs mobile phone charger by utilizing 99% methanol. Because less methanol was required to produce the same quantity of electricity, less methanol storage was required [11,71]. Toshiba unveiled the Dynario DMFC power pack, which can supply up to 2 W of power to a digital device. Panasonic created a stack of high-power DMFCs with an average power output of 20 W that can run for 5,000 hours. Smart Fuel Cell developed a portable DMFC system to power electrical and electronic devices on military vehicles. This system has advantages over traditional battery or diesel-powered systems, including low sound emission, long operational time, lower maintenance requirements, and lighter weight due to elimination of the heavy diesel generator [25,27]. IdaTech has developed the use of reformed methanol fuel cells (RMFCs) as a backup power supply for telecommunication base stations in the US. IdaTech stated that it can run for forty hours with a 5 kW output and 50 litres of a dilute methanol solution. More than 500 DMFC-based telecom devices have been installed worldwide [10,12]. DMFCs can also function as backup power systems for an uninterruptible emergency power supply for security needs, such as a police radio station and fire alarm system. The Yorkshire Ambulance Service in the UK uses DMFC power supply to charge ambulance equipment. It ensures that the equipment may be utilized at any moment, including when the ambulance breaks down, potentially saving someone's life. In Denmark, a hearing aid based on DMFCs was developed, as illustrated in Figure 7.9a. It can work continuously for 24 hours using only 200 mL of methanol fuel [87]. In the industrial field, Oorja Protonics (Figure 7.9b) had pioneered the development and commercialization of a DMFCs system for the onboard battery charging of material handling vehicles.

NDC Power developed the EOS DEFCs to generate power in the 1e250 W range for DEFC applications. The EOS DEFCs was used as a power pack to recharge portable devices, such as laptop computers and mobile phones.

FIGURE 7.9
(a) DMFCs-operated hearing aid, (b) Oorja's DMFCs for material handling vehicle, and (c) prototype car running on ethanol as fuel.

Source: Reproduced with permission from [87].

During the Shell Ecomarathon Asia 2012, a Thai team created a very efficient ethanol-powered prototype automobile that could drive 2903 kilometers on just 1 L of ethanol, as seen in Figure 7.9c. This demonstrated that ethanol may be a fuel with considerable potential for usage in transportation in the future. Horizon had developed an education kit applying ethanol as a fuel, known as the Bio-Energy Discovery Kit [87].

7.4 DLFC for Electricity Generation

DLFCs are electrochemical devices that use oxidation and reduction processes to directly transform the chemical energy of liquid fuels into electrical energy. Direct liquid fuel cells (DLFCs) are a promising type of fuel cell due to its high energy density, simple structure, tiny fuel cartridge, quick recharge, and ease of storage and transport. The most commonly used fuels were alcohols (methanol and ethanol), but glycols and acids were also used. DLFCs are thought to be the best options for highly specialized military applications and mid-size outputs between a few hundred watts to about 3 kW. It is possible for DLFCs to function without an electrical power source [11,37]. Furthermore, the energy density of liquid fuels is typically higher than that of lithium-ion polymers, at about 600 Wh kg^{-1}. This indicates that, in contrast to rechargeable batteries, which require hours to charge when they run low, DLFCs can offer longer mobile phone conversation times, longer laptop computer usage times, more power available to meet consumer demand, and convenient instantaneous refuelling. In DLFCs, liquid fuel is delivered directly to the anode for oxidation, while air or oxygen gas is sent to the cathode for reduction. The charged ions move through the electrolyte, while electrons travel through an external circuit. The anode produces certain side products due to incomplete fuel oxidation, while the cathode mostly produces water [52,53].

7.5 Conclusion

The increasing sophistication of DLFC technology shows great potential in meeting the worldwide need for environmentally friendly and economically viable energy options. The chapter offers a comprehensive review of the important factors that govern the development, improvement, and commercialization of DLFCs. This technology has the ability to transform energy systems in several industries by connecting scientific advancements with real-world applications. System designers have worked tirelessly to make DLFCs more efficient, long-lasting, and scalable throughout their development.

Engineering and material science have made great strides in solving important problems, including fuel crossover, optimizing electrode materials, and integrating systems. New membrane materials, compact system designs, and high-performance catalysts have greatly improved DLFC systems' energy density and operational stability. Furthermore, DLFCs may be easily modified for use in a wide variety of contexts due to its modular nature. These contexts can include anything from small electronic devices to massive energy grids.

The broad adoption of DLFC technology is heavily dependent on its commercialization. High manufacturing costs, restricted fuel supply, and rivalry from proven energy technologies have slowed the progress from lab to market, despite the remarkable technological improvements. To overcome these obstacles, a multi-pronged strategy is required, as this chapter explains. Promoting innovation and building strong supply chains for DLFC fuels and components requires concerted action from lawmakers, industry stakeholders, and academics.

Due to its advantages, both financial and ecological, DLFCs are quickly becoming an integral part of the worldwide movement towards more sustainable energy sources. By decreasing reliance on fossil fuels and emissions of greenhouse gases, the use of liquid fuels like methanol and ethanol, which are produced from renewable resources, is in line with worldwide sustainability objectives. Moreover, DLFC systems are more appealing to underserved and faraway areas because of the possibility of localized fuel generation, which promotes energy independence and decentralization. Improving fuel-processing procedures, decreasing prices, and increasing the operating lifetime of DLFC systems should be the focus of future research as the technology develops. The acceleration of commercialization will also depend on the standardization of testing techniques and the creation of worldwide performance and safety standards. Also, getting the word out about how great DLFC technology is will be key to getting people to buy it and use it. Finally, the revolutionary possibilities of direct liquid fuel cell technology are highlighted by its commercialization and system design components in the energy sector. New opportunities for DLFCs will be unlocked as long as creative research, supporting regulations, and market-driven strategies continue to converge, despite the fact that hurdles will always be there. With further work in these areas, DLFC technology has a great chance of becoming an essential component of long-term sustainable energy solutions.

References

[1] Q. Hassan, P. Viktor, T.J. Al-Musawi, B. Mahmood Ali, S. Algburi, H.M. Alzoubi, A. Khudhair Al-Jiboory, A. Zuhair Sameen, H.M. Salman, M. Jaszczur, The renewable energy role in the global energy Transformations, *Renew. Energy Focus.* 48 (2024) 100545. https://doi.org/10.1016/J.REF.2024.100545.

[2] A.D.A. Bin Abu Sofian, H.R. Lim, H. Siti Halimatul Munawaroh, Z. Ma, K.W. Chew, P.L. Show, Machine learning and the renewable energy revolution: Exploring solar and wind energy solutions for a sustainable future including innovations in energy storage, *Sustain. Dev.* 32 (2024) 3953–3978. https://doi.org/10.1002/SD.2885.

[3] E.T. Sayed, A.G. Olabi, K. Elsaid, M. Al Radi, C. Semeraro, M.H. Doranehgard, M.E. Eltayeb, M.A. Abdelkareem, Application of artificial intelligence techniques for modeling, optimizing, and controlling desalination systems powered by renewable energy resources, *J. Clean. Prod.* 413 (2023) 137486. https://doi.org/10.1016/J.JCLEPRO.2023.137486.

[4] E.T. Sayed, A.G. Olabi, K. Elsaid, M. Al Radi, R. Alqadi, M. Ali Abdelkareem, Recent progress in renewable energy based-desalination in the Middle East and North Africa MENA region, *J. Adv. Res.* 48 (2023) 125–156. https://doi.org/10.1016/J.JARE.2022.08.016.

[5] N.A.A. Qasem, G.A.Q. Abdulrahman, A recent comprehensive review of fuel cells: History, types, and applications, *Int. J. Energy Res.* 2024 (2024) 7271748. https://doi.org/10.1155/2024/7271748.

[6] Y. Luo, Y. Wu, B. Li, T. Mo, Y. Li, S.P. Feng, J. Qu, P.K. Chu, Development and application of fuel cells in the automobile industry, *J. Energy Storage.* 42 (2021) 103124. https://doi.org/10.1016/J.EST.2021.103124.

[7] I. Staffell, D. Scamman, A. Velazquez Abad, P. Balcombe, P.E. Dodds, P. Ekins, N. Shah, K.R. Ward, The role of hydrogen and fuel cells in the global energy system, *Energy Environ. Sci.* 12 (2019) 463–491. https://doi.org/10.1039/C8EE01157E.

[8] A. Jamil, S. Rafiq, T. Iqbal, H.A.A. Khan, H.M. Khan, B. Azeem, M.Z. Mustafa, A.S. Hanbazazah, Current status and future perspectives of proton exchange membranes for hydrogen fuel cells, *Chemosphere.* 303 (2022) 135204. https://doi.org/10.1016/J.CHEMOSPHERE.2022.135204.

[9] R. Rath, P. Kumar, S. Mohanty, S.K. Nayak, Recent advances, unsolved deficiencies, and future perspectives of hydrogen fuel cells in transportation and portable sectors, *Int. J. Energy Res.* 43 (2019) 8931–8955. https://doi.org/10.1002/ER.4795.

[10] N. Sazali, W.N.W. Salleh, A.S. Jamaludin, M.N.M. Razali, New perspectives on fuel cell technology: A brief review, *Membranes.* 10 (2020) 99. https://doi.org/10.3390/MEMBRANES10050099.

[11] N. Shaari, S.K. Kamarudin, R. Bahru, S.H. Osman, N.A.I. Md Ishak, Progress and challenges: Review for direct liquid fuel cell, *Int. J. Energy Res.* 45 (2021) 6644–6688. https://doi.org/10.1002/ER.6353.

[12] Ş. Neaţu, F. Neaţu, I. M. Chirica, I. Borbáth, E. Tálas, A. Tompos, S. Somacescu, P. Osiceanu, M. A. Folgado, A. M. Chaparro, M. Florea, Recent progress in electrocatalysts and electrodes for portable fuel cells, *J. Mater. Chem. A.* 9 (2021) 17065–17128. https://doi.org/10.1039/D1TA03644K.

[13] H. Huang, X. Guo, C. Zhang, L. Yang, Q. Jiang, H. He, M.A. Amin, W.A. Alshahrani, J. Zhang, X. Xu, Y. Yamauchi, Advancements in noble metal-decorated porous carbon nanoarchitectures: Key catalysts for direct liquid fuel cells, *ACS Nano.* 18 (2024) 10341–10373. https://doi.org/10.1021/ACSNANO.3C08486/ASSET/IMAGES/MEDIUM/NN3C08486_0021.GIF.

[14] J. Liu, Z. Zhou, B. Yue, Z. Sun, Z. Sun, Chemical looping induced CH3OH–H2-PEMFC scheme for fuel cell vehicle: Parameter optimization and feasibility analysis, *J. Power Sources.* 479 (2020) 228790. https://doi.org/10.1016/J.JPOWSOUR.2020.228790.

[15] C.P. Chang, Y.C. Wu, W.Y. Chen, C. Pan, Y.C. Su, Y.J. Huang, F.G. Tseng, A hybrid phosphorus-acid fuel cell system incorporated with oxidative steam reforming of methanol (OSRM) reformer, *Renew. Energy*. 153 (2020) 530–538. https://doi.org/10.1016/J.RENENE.2020.01.137.
[16] S.H. Osman, O.H. Chyuan, S.K. Kamarudin, N. Shaari, I.H. Hanapi, Z. Zakaria, N.H. Ahmad Zaidi, S.A. Adnan, R. Bahru, Nanocatalysts in direct liquid fuel cells: Advancements for superior performance and energy sustainability, *Int. J. Green Energy*. (2024). https://doi.org/10.1080/15435075.2024.2396070.
[17] A.G. Olabi, T. Wilberforce, M.A. Abdelkareem, Fuel cell application in the automotive industry and future perspective, *Energy*. 214 (2021) 118955. https://doi.org/10.1016/J.ENERGY.2020.118955.
[18] A. Alaswad, A. Omran, J.R. Sodre, T. Wilberforce, G. Pignatelli, M. Dassisti, A. Baroutaji, A.G. Olabi, Technical and commercial challenges of Proton-Exchange Membrane (PEM) fuel cells, *Energies 2021*. 14 (2020) 144. https://doi.org/10.3390/EN14010144.
[19] A.A. Wani, N. Shaari, S.K. Kamarudin, N.F. Raduwan, Y.N. Yusoff, A.M. Khan, S. Yousuf, M.N.M. Ansari, Critical review on composite-based polymer electrolyte membranes toward fuel cell applications: Progress and perspectives, *Energy and Fuels*. (2024). https://doi.org/10.1021/ACS.ENERGYFUELS.4C02516/ASSET/IMAGES/MEDIUM/EF4C02516_0014.GIF.
[20] M. Tanveer, T. Ambreen, H. Khan, G.M. Kim, C.W. Park, Paper-based microfluidic fuel cells and their applications: A prospective review, *Energy Convers. Manag*. 264 (2022) 115732. https://doi.org/10.1016/j.enconman.2022.115732.
[21] A. Singh, A. Gangwar, S. Kumar, Prospects and limitations of existing biofuels and emerging trends in the utilization of nanoparticles for enhanced biofuel production and microbial fuel cell efficiency, *Biofuels*. (2024). https://doi.org/10.1080/17597269.2024.2431766.
[22] O. Emmanuel, Rozina, T.C. Ezeji, Utilization of biomass-based resources for biofuel production: A mitigating approach towards zero emission, *Sustain. Chem. One World*. 2 (2024) 100007. https://doi.org/10.1016/J.SCOWO.2024.100007.
[23] M. Rumayor, A. Dominguez-Ramos, A. Irabien, Formic acid manufacture: Carbon dioxide utilization alternatives, *Appl. Sci*. 8 (2018) 914. https://doi.org/10.3390/APP8060914.
[24] I. Dutta, S. Chatterjee, H. Cheng, R.K. Parsapur, Z. Liu, Z. Li, E. Ye, H. Kawanami, J.S.C. Low, Z. Lai, X.J. Loh, K.W. Huang, Formic acid to power towards low-carbon economy, *Adv. Energy Mater*. 12 (2022) 2103799. https://doi.org/10.1002/AENM.202103799.
[25] R. Bhaskaran, B.G. Abraham, R. Chetty, Recent advances in electrocatalysts, mechanism, and cell architecture for direct formic acid fuel cells, *Wiley Interdiscip. Rev. Energy Environ*. 11 (2022) e419. https://doi.org/10.1002/WENE.419.
[26] G. Amba Prasad Rao, K.J. Reddy, R.M. Reddy, K. Madhu Murthy, G. Naga Srinivasulu, Direct methanol fuel cells for automotive applications: A review, *Int. J. Ambient Energy*. 43 (2022) 7349–7370. https://doi.org/10.1080/01430750.2022.2063380.
[27] A. Gisario, M. Kazarian, F. Martina, M. Mehrpouya, Metal additive manufacturing in the commercial aviation industry: A review, *J. Manuf. Syst*. 53 (2019) 124–149. https://doi.org/10.1016/J.JMSY.2019.08.005.

[28] S.S. Kamble, A. Gunasekaran, A. Ghadge, R. Raut, A performance measurement system for industry 4.0 enabled smart manufacturing system in SMMEs-A review and empirical investigation, *Int. J. Prod. Econ.* 229 (2020) 107853. https://doi.org/10.1016/J.IJPE.2020.107853.

[29] T.L. Nguyen, Q.N. Tran, P.K.T. Nguyen, Mathematical modeling of anion exchange membrane-based direct urea/O2 fuel cell incorporating the effects of under-rib mass transport and gaseous bubble formation, *J. Electroanal. Chem.* 945 (2023) 117682. https://doi.org/10.1016/J.JELECHEM.2023.117682.

[30] F.M. Souza, V.S. Pinheiro, T.C. Gentil, L.E.B. Lucchetti, J.C.M. Silva, M.L.M.G. Santos, I. De Oliveira, W.M.C. Dourado, G. Amaral-Labat, S. Okamoto, M.C. Santos, Alkaline direct liquid fuel cells: Advances, challenges and perspectives, *J. Electroanal. Chem.* 922 (2022) 116712. https://doi.org/10.1016/J.JELECHEM.2022.116712.

[31] V. Narayanamurthy, Z.E. Jeroish, K.S. Bhuvaneshwari, P. Bayat, R. Premkumar, F. Samsuri, M.M. Yusoff, Advances in passively driven microfluidics and lab-on-chip devices: A comprehensive literature review and patent analysis, *RSC Adv.* 10 (2020) 11652–11680. https://doi.org/10.1039/D0RA00263A.

[32] H. Shabgard, X. Li, A. Faghri, Opportunities and challenges in passive thermal-fluid and energy systems, *ASME J. Heat Mass Transf.* 145 (2023). https://doi.org/10.1115/1.4055342/1145753.

[33] I.A. Hassan, H.S. Ramadan, M.A. Saleh, D. Hissel, Hydrogen storage technologies for stationary and mobile applications: Review, analysis and perspectives, *Renew. Sustain. Energy Rev.* 149 (2021) 111311. https://doi.org/10.1016/J.RSER.2021.111311.

[34] R.A. Felseghi, E. Carcadea, M.S. Raboaca, C.N. Trufin, C. Filote, Hydrogen fuel cell technology for the sustainable future of stationary applications, *Energies.* 12 (2019) 4593. https://doi.org/10.3390/EN12234593.

[35] V. Cigolotti, M. Genovese, P. Fragiacomo, Comprehensive review on fuel cell technology for stationary applications as sustainable and efficient poly-generation energy systems, *Energies.* 14 (2021) 4963. https://doi.org/10.3390/EN14164963.

[36] Z. Ma, U. Legrand, E. Pahija, J.R. Tavares, D.C. Boffito, From CO2to formic acid fuel cells, *Ind. Eng. Chem. Res.* 60 (2021) 803–815. https://doi.org/10.1021/ACS.IECR.0C04711/ASSET/IMAGES/MEDIUM/IE0C04711_0008.GIF.

[37] A.M. Othman, A.A. El-Fergany, Optimal dynamic operation and modeling of parallel connected multi-stacks fuel cells with improved slime mould algorithm, *Renew. Energy.* 175 (2021) 770–782. https://doi.org/10.1016/J.RENENE.2021.04.148.

[38] D. Ma, W. Chen, X. Ruan, A review of voltage/current sharing techniques for series-parallel-connected modular power conversion systems, *IEEE Trans. Power Electron.* 35 (2020) 12383–12400. https://doi.org/10.1109/TPEL.2020.2984714.

[39] W. Qian, D.P. Wilkinson, J. Shen, H. Wang, J. Zhang, Architecture for portable direct liquid fuel cells, *J. Power Sources.* 154 (2006) 202–213. https://doi.org/10.1016/J.JPOWSOUR.2005.12.019.

[40] Z. Pan, F. Xie, Z. Zhang, Z. Zhao, L. Wu, W. Li, Y. Zhu, X. Huo, Y. Liu, X. Zhang, R. Chen, L. An, Overall design of a gradient-ordered membrane electrode assembly for direct liquid fuel cells, *Adv. Funct. Mater.* 34 (2024) 2404710. https://doi.org/10.1002/ADFM.202404710.

[41] Z. Liu, J. Chen, H. Liu, C. Yan, Y. Hou, Q. He, J. Zhang, D. Hissel, Anode purge management for hydrogen utilization and stack durability improvement of PEM fuel cell systems, Appl. *Energy*. 275 (2020) 115110. https://doi.org/10.1016/J.APENERGY.2020.115110.

[42] J.C. Kurnia, A.P. Sasmito, T. Shamim, Advances in proton exchange membrane fuel cell with dead-end anode operation: A review, *Appl. Energy*. 252 (2019) 113416. https://doi.org/10.1016/J.APENERGY.2019.113416.

[43] X. Yan, C. Guan, Y. Zhang, K. Jiang, G. Wei, X. Cheng, S. Shen, J. Zhang, Flow field design with 3D geometry for proton exchange membrane fuel cells, *Appl. Therm. Eng.* 147 (2019) 1107–1114. https://doi.org/10.1016/J.APPLTHERMALENG.2018.09.110.

[44] K. Xiong, W. Wu, S. Wang, L. Zhang, Modeling, design, materials and fabrication of bipolar plates for proton exchange membrane fuel cell: A review, *Appl. Energy*. 301 (2021) 117443. https://doi.org/10.1016/J.APENERGY.2021.117443.

[45] M. Sauermoser, N. Kizilova, B.G. Pollet, S. Kjelstrup, Flow field patterns for proton exchange membrane fuel cells, *Front. Energy Res.* 8 (2020) 482555. https://doi.org/10.3389/FENRG.2020.00013/BIBTEX.

[46] Y. Wang, X. Liao, G. Liu, H. Xu, C. Guan, H. Wang, H. Li, W. He, Y. Qin, Review of flow field designs for polymer electrolyte membrane fuel cells, *Energies*. 16 (2023) 4207. https://doi.org/10.3390/EN16104207.

[47] R. Gundlapalli, S. Jayanti, Effective splitting of serpentine flow field for applications in large-scale flow batteries, *J. Power Sources*. 487 (2021) 229409. https://doi.org/10.1016/J.JPOWSOUR.2020.229409.

[48] M.A. Babay, M. Adar, A. Chebak, M. Mabrouki, Comparative sustainability analysis of serpentine flow-field and straight channel PEM fuel cell designs, *Int. J. Syst. Assur. Eng. Manag.* 15 (2024) 3954–3970. https://doi.org/10.1007/S13198-024-02395-8/FIGURES/15.

[49] B.H. Lim, E.H. Majlan, A. Tajuddin, T. Husaini, W.R. Wan Daud, N.A. Mohd Radzuan, M.A. Haque, Comparison of catalyst-coated membranes and catalyst-coated substrate for PEMFC membrane electrode assembly: A review, *Chinese J. Chem. Eng.* 33 (2021) 1–16. https://doi.org/10.1016/J.CJCHE.2020.07.044.

[50] M. Chen, C. Zhao, F. Sun, J. Fan, H. Li, H. Wang, Research progress of catalyst layer and interlayer interface structures in membrane electrode assembly (MEA) for proton exchange membrane fuel cell (PEMFC) system, *eTransportation*. 5 (2020) 100075. https://doi.org/10.1016/J.ETRAN.2020.100075.

[51] Q. Ye, X. Zhao, R. Jin, F. Dong, H. Xie, B. Deng, Advances and challenges in membrane electrode assembly electrolyzers for CO2 reduction, *J. Mater. Chem. A*. 11 (2023) 21498–21515. https://doi.org/10.1039/D3TA03757F.

[52] M. Bekmezci, R. Bayat, M. Akina, H. Burhan, I. Isik, F. Şen, The electrocatalysts with pH of the electrolyte for the complete pathways of the oxidation reactions, *Nanomater. Direct Alcohol Fuel Cells Charact. Des. Electrocatal.* (2021) 95–107. https://doi.org/10.1016/B978-0-12-821713-9.00020-2.

[53] K.K. Maniam, R. Chetty, R. Thimmappa, S. Paul, Progress in the development of electrodeposited catalysts for direct liquid fuel cell applications, *Appl. Sci.* 12 (2022) 501. https://doi.org/10.3390/APP12010501.

[54] T.A.M. Suter, K. Smith, J. Hack, L. Rasha, Z. Rana, G.M.A. Angel, P.R. Shearing, T.S. Miller, D.J.L. Brett, Engineering catalyst layers for next-generation polymer electrolyte fuel cells: A review of design, materials, and methods, *Adv. Energy Mater.* 11 (2021) 2101025. https://doi.org/10.1002/AENM.202101025.

[55] H. Chen, Z. Wu, M. Zheng, T. Liu, C. Yan, J. Lu, S. Zhang, Catalytic materials for lithium-sulfur batteries: Mechanisms, design strategies and future perspective, *Mater. Today.* 52 (2022) 364–388. https://doi.org/10.1016/J.MATTOD.2021.10.026.
[56] C. Molochas, P. Tsiakaras, Carbon monoxide tolerant Pt-based electrocatalysts for H2-PEMFC applications: Current progress and challenges, *Catalysts.* 11 (2021) 1127. https://doi.org/10.3390/CATAL11091127.
[57] V.F. Valdés-López, T. Mason, P.R. Shearing, D.J.L. Brett, Carbon monoxide poisoning and mitigation strategies for polymer electrolyte membrane fuel cells—A review, *Prog. Energy Combust. Sci.* 79 (2020) 100842. https://doi.org/10.1016/J.PECS.2020.100842.
[58] J. Zhou, J. Cao, Y. Zhang, J. Liu, J. Chen, M. Li, W. Wang, X. Liu, Overcoming undesired fuel crossover: Goals of methanol-resistant modification of polymer electrolyte membranes, *Renew. Sustain. Energy Rev.* 138 (2021) 110660. https://doi.org/10.1016/J.RSER.2020.110660.
[59] X. Li, Z. Miao, L. Marten, I. Blankenau, Experimental measurements of fuel and water crossover in an active DMFC, *Int. J. Hydrogen Energy.* 46 (2021) 4437–4446. https://doi.org/10.1016/J.IJHYDENE.2020.11.027.
[60] M.E. Abdelrahman, H. Zhang, G. Wu, X. Li, S. Litster, Half-cell electrode assessments of a crossover-tolerant direct methanol fuel cell with a platinum group metal-free cathode, *Electrochim. Acta.* 416 (2022) 140262. https://doi.org/10.1016/J.ELECTACTA.2022.140262.
[61] I. Vincent, E.C. Lee, H.M. Kim, Solutions to the water flooding problem for unitized regenerative fuel cells: Status and perspectives, *RSC Adv.* 10 (2020) 16844–16860. https://doi.org/10.1039/D0RA00434K.
[62] O.S. Ijaodola, Z. El- Hassan, E. Ogungbemi, F.N. Khatib, T. Wilberforce, J. Thompson, A.G. Olabi, Energy efficiency improvements by investigating the water flooding management on proton exchange membrane fuel cell (PEMFC), *Energy.* 179 (2019) 246–267. https://doi.org/10.1016/J.ENERGY.2019.04.074.
[63] Y. Pan, H. Wang, N.P. Brandon, Gas diffusion layer degradation in proton exchange membrane fuel cells: Mechanisms, characterization techniques and modelling approaches, *J. Power Sources.* 513 (2021) 230560. https://doi.org/10.1016/J.JPOWSOUR.2021.230560.
[64] Y. Wu, H. Rabiee, X.S. Zhao, G. Wang, Y. Jiang, Insights into electrolyte flooding in flexible gas diffusion electrodes for CO 2 electrolysis: From mechanisms to effective mitigation strategies, *J. Mater. Chem. A.* 12 (2024) 14206–14228. https://doi.org/10.1039/D4TA01994F.
[65] S. Ott, A. Orfanidi, H. Schmies, B. Anke, H.N. Nong, J. Hübner, U. Gernert, M. Gliech, M. Lerch, P. Strasser, Ionomer distribution control in porous carbon-supported catalyst layers for high-power and low Pt-loaded proton exchange membrane fuel cells, *Nat. Mater.* 19 (2019) 77–85. https://doi.org/10.1038/s41563-019-0487-0.
[66] P.C. Okonkwo, C. Otor, A review of gas diffusion layer properties and water management in proton exchange membrane fuel cell system, *Int. J. Energy Res.* 45 (2021) 3780–3800. https://doi.org/10.1002/ER.6227.
[67] J. Zhang, B. Wang, J. Jin, S. Yang, G. Li, A review of the microporous layer in proton exchange membrane fuel cells: Materials and structural designs based on water transport mechanism, Renew. *Sustain. Energy Rev.* 156 (2022) 111998. https://doi.org/10.1016/J.RSER.2021.111998.

[68] S. Manikandan, R. Subbaiya, M. Saravanan, M. Ponraj, M. Selvam, A. Pugazhendhi, A critical review of advanced nanotechnology and hybrid membrane based water recycling, reuse, and wastewater treatment processes, *Chemosphere*. 289 (2022) 132867. https://doi.org/10.1016/J.CHEMOSPHERE.2021.132867.

[69] B.P. Chaplin, The prospect of electrochemical technologies advancing worldwide water treatment, *Acc. Chem. Res*. 52 (2019) 596–604. https://doi.org/10.1021/ACS.ACCOUNTS.8B00611/SUPPL_FILE/AR8B00611_SI_001.PDF.

[70] R. Roy, C.M. Lampert, I. Stoyneva, When dinosaurs fly: The role of firm capabilities in the 'avianization' of incumbents during disruptive technological change, *Strateg. Entrep. J*. 12 (2018) 261–284. https://doi.org/10.1002/SEJ.1278.

[71] Y. Ito, Interorganizational business development utilizing legitimacy for resource mobilization in large firms: Successful and unsuccessful cases, *Ind. Mark. Manag*. 75 (2018) 80–89. https://doi.org/10.1016/J.INDMARMAN.2018.04.002.

[72] O.S. Popel ', A.B. Tarasenko, S.P. Filippov, Fuel cell based power-generating installations: State of the art and future prospects, *Therm. Eng*. 65 (2018) 859–874. https://doi.org/10.1134/S0040601518120078/METRICS.

[73] Y. Wang, D.F. Ruiz Diaz, K.S. Chen, Z. Wang, X.C. Adroher, Materials, technological status, and fundamentals of PEM fuel cells—A review, *Mater. Today*. 32 (2020) 178–203. https://doi.org/10.1016/J.MATTOD.2019.06.005.

[74] M. Singh, M.K. Singla, M. Safaraliev, K. Singh, I. Odinaev, A. Abdel Menaem, Advancements and challenges of fuel cell integration in electric vehicles: A comprehensive analysis, *Int. J. Hydrogen Energy*. 88 (2024) 1386–1397. https://doi.org/10.1016/J.IJHYDENE.2024.09.212.

[75] X. Yu, A. Manthiram, Scalable membraneless direct liquid fuel cells based on a catalyst-selective strategy, *Energy Environ. Mater*. 1 (2018) 13–19. https://doi.org/10.1002/EEM2.12000.

[76] J. Chang, G. Wang, W. Zhang, Y. Yang, Atomically dispersed catalysts for small molecule electrooxidation in direct liquid fuel cells, *J. Energy Chem*. 68 (2022) 439–453. https://doi.org/10.1016/J.JECHEM.2021.12.017.

[77] Z.A.C. Ramli, S.K. Kamarudin, Platinum-based catalysts on various carbon supports and conducting polymers for direct methanol fuel cell applications: A review, *Nanoscale Res. Lett*. 13 (2018) 1–25. https://doi.org/10.1186/S11671-018-2799-4.

[78] Z.A.C. Ramli, J. Pasupuleti, S.K. Kamarudin, A.M. Zainoodin, W.N.R.W. Isahak, S.P. Koh, S.T. Kiong, Harnessing the potential of hollow graphitic carbon nanocages for enhanced methanol oxidation using PtRu nanoparticles, *Polym*. 16 (2024) 2684. https://doi.org/10.3390/POLYM16192684.

[79] D. Zengel, M. Stehle, O. Deutschmann, M. Casapu, J.D. Grunwaldt, Impact of gas phase reactions and catalyst poisons on the NH3-SCR activity of a V2O5-WO3/TiO2 catalyst at pre-turbine position, *Appl. Catal. B Environ*. 288 (2021) 119991. https://doi.org/10.1016/J.APCATB.2021.119991.

[80] N. Shaari, Z. Zakaria, S.K. Kamarudin, The optimization performance of cross-linked sodium alginate polymer electrolyte bio-membranes in passive direct methanol/ethanol fuel cells, *Int. J. Energy Res*. 43 (2019) 8275–8285. https://doi.org/10.1002/ER.4825.

[81] M. Mazurkiewicz-Pawlicka, A. Malolepszy, A. Mikolajczuk-Zychora, B. Mierzwa, A. Borodzinski, L. Stobinski, A simple method for enhancing the catalytic activity of Pd deposited on carbon nanotubes used in direct formic acid fuel cells, *Appl. Surf. Sci*. 476 (2019) 806–814. https://doi.org/10.1016/J.APSUSC.2019.01.114.

[82] J. Zhang, L. Wu, L. Xu, D. Sun, H. Sun, Y. Tang, Recent advances in phosphorus containing noble metal electrocatalysts for direct liquid fuel cells, *Nanoscale*. 13 (2021) 16052–16069. https://doi.org/10.1039/D1NR04218A.
[83] M.T.H. Siddiqui, S. Nizamuddin, H.A. Baloch, N.M. Mubarak, M. Al-Ali, S.A. Mazari, A.W. Bhutto, R. Abro, M. Srinivasan, G. Griffin, Fabrication of advance magnetic carbon nano-materials and their potential applications: A review, *J. Environ. Chem. Eng.* 7 (2019) 102812. https://doi.org/10.1016/J.JECE.2018.102812.
[84] O.S.J. Elham, S.K. Kamarudin, N. Shaari, A.M. Zainoodin, Z. Zakaria, M.R. Yusof, Development of low-cost nafion-lignin composite conductive membranes for application in direct methanol fuel cells, *J. Environ. Chem. Eng.* 12 (2024) 111514. https://doi.org/10.1016/J.JECE.2023.111514.
[85] M. Ahmaruzzaman, S.R. Mishra, V. Gadore, G. Yadav, S. Roy, B. Bhattacharjee, A. Bhuyan, B. Hazarika, J. Darabdhara, K. Kumari, Phenolic compounds in water: From toxicity and source to sustainable solutions—An integrated review of removal methods, advanced technologies, cost analysis, and future prospects, *J. Environ. Chem. Eng.* 12 (2024) 112964. https://doi.org/10.1016/J.JECE.2024.112964.
[86] Y. Yang, S. Yuan, J. Liu, Z. Zhang, T. Lu, Effect of catalyst ink particle size on the structure of the catalyst layer and electrical performance in the process of ultrasonic spray manufacturing PEMFCs, *Energy*. 294 (2024) 130605. https://doi.org/10.1016/J.ENERGY.2024.130605.
[87] B.C. Ong, S.K. Kamarudin, S. Basri, Direct liquid fuel cells: A review, *Int. J. Hydrogen Energy*. 42 (2017) 10142–10157. https://doi.org/10.1016/J.IJHYDENE.2017.01.117.
[88] F. Bienen, D. Kopljar, A. Löwe, P. Aßmann, M. Stoll, P. Rößner, N. Wagner, A. Friedrich, E. Klemm, Utilizing formate as an energy carrier by coupling CO2 electrolysis with fuel cell devices, *Chemie Ing. Tech.* 91 (2019) 872–882. https://doi.org/10.1002/CITE.201800212.
[89] D.M. Fadzillah, S.K. Kamarudin, M.A. Zainoodin, M.S. Masdar, Critical challenges in the system development of direct alcohol fuel cells as portable power supplies: An overview, *Int. J. Hydrogen Energy*. 44 (2019) 3031–3054. https://doi.org/10.1016/J.IJHYDENE.2018.11.089.
[90] F. Al Mubarak, R. Rezaee, D.A. Wood, Economic, societal, and environmental impacts of available energy sources: A review, *Eng.* 5 (2024) 1232–1265. https://doi.org/10.3390/ENG5030067.
[91] V. Raj, Direct methanol fuel cells in portable applications: Materials, designs, operating parameters, and practical steps toward commercialization, *Direct Methanol Fuel Cell Technol.* (2020) 495–525. https://doi.org/10.1016/B978-0-12-819158-3.00016-1.

8

Conclusion and Future Challenges in Direct Liquid Fuel Cells

Norazuwana Shaari, Siti Hasanah Osman, and Ajaz Ahmad Wani

8.1 Introduction

DLFCs, which have a number of advantages over conventional hydrogen-based fuel cells and traditional batteries, have become a promising technology in the field of fuel cells [1]. Unlike traditional fuel cells, which require hydrogen, DLFCs use liquid fuels directly, such as methanol, ethanol, or hydrocarbons, negating the need for complicated and expensive hydrogen delivery and storage infrastructure. Due to their special characteristics, DLFCs are adaptable and have a broad variety of possible applications in numerous industries. From this perspective, this section will examine the findings reached from the prospective uses of DLFCs and the difficulties that must be overcome in order to fully actualize their widespread acceptance and integration into our energy system. Dermici and team [2] investigated several DLFC types concentrated on thermodynamic data and biologically hazardous risks associated with chemicals employed as liquid fuels. Soloveichik et al. [3] examined the differences between organic fuel and inorganic fuel in a study on liquid fuel cells. By allowing the direct use of liquid fuels without the requirement for preceding hydrogen reforming, DLFCs fundamentally offer a ground-breaking method of energy conversion. Due to their distinct feature, DLFCs stand apart from standard hydrogen-based fuel cells and traditional batteries, making them an attractive option to address a variety of energy-related problems.

The widespread use and commercialization of direct liquid fuel cells face a number of obstacles. A big obstacle is raising DLFCs' general efficiency. Although they have a higher energy density than standard batteries, contemporary DLFCs might be less efficient than hydrogen fuel cells. Efficiency improvements for DLFCs would make them more competitive as a clean and sustainable energy source. The creation of long-lasting and

 DOI: 10.1201/9781003512592-8

affordable electrode materials is another major obstacle. The durability and efficiency of the fuel cells can be impacted by the electrochemical processes in DLFCs over time, which can result in catalyst poisoning and deterioration. Finding robust electrode materials that can resist the challenging operating conditions and deliver long-lasting performance must be a top priority for researchers and industry partners. Due to potential incomplete oxidation, the side product of liquid fuel oxidation is produced in the anode, but the cathode's waste product is water. Figure 8.1 depicts the DLFC method in detail.

An additional obstacle to go beyond is fuel crossover. Liquid fuels may transfer from one electrode to the other while the device is running, reducing efficiency and raising the possibility of fuel waste. To improve energy conversion and ensure effective use of liquid fuels, fuel crossing must be minimized [3]. In order to facilitate the widespread deployment of DLFCs, it is also necessary to create and improve the infrastructure for liquid fuel distribution and refilling. The majority of the current fueling infrastructure is geared around traditional internal combustion engines, so adding new DLFC refuelling stations to the network or integrating existing ones will need substantial financial outlays and careful planning. Additionally, concerns with safety pertaining to the handling and storage of liquid fuels must be fully addressed. It is essential to assure the secure handling, keeping, and use of liquid fuels in DLFCs in order to avoid mishaps and guarantee the general public's approval of this technology.

The field of distributed energy generation is another area where DLFCs' adaptability is further enhanced. DLFCs provide a consistent and effective

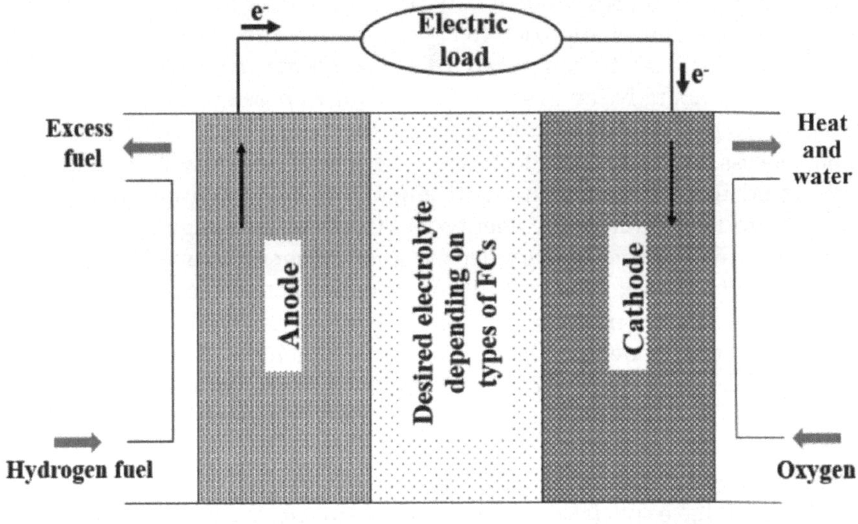

FIGURE 8.1
Reaction in the direct liquid fuel cell system.

power source in remote or off-grid areas, completing intermittent renewable energy systems like solar and wind [4]. The development of energy hubs that can sustain regional communities and enterprises is made possible by the simplicity of storing and transporting liquid fuels. The use of DLFCs to power unmanned aerial vehicles (UAVs) and electric vehicles (EVs) could improve the transportation industry. The capability of DLFCs to offer rapid refuelling and increased range fits well with the changing requirements of sustainable mobility and aerial operations.

As a flexible and revolutionary energy solution for a variety of industries, direct liquid fuel cells hold enormous promise. Although their direct reliance on liquid fuels has many benefits, the full potential of DLFCs must be realized in order to lead the way for a cleaner and more sustainable energy future. These issues include those relating to efficiency, durability, fuel crossover, infrastructure, and safety [5,6]. To overcome these obstacles and promote the general implementation of DLFCs in the upcoming years, researchers, companies, and governments must work together. This article is divided into several sections: the introduction; the several conflict in direct fluid fuel cells, which include costing; the DLFC class that comes before each category of accomplishments and developments that have been discussed; the current fuel cell application; and finally, the conclusion.

8.2 Advantages and Impact of DLFCs

DLFCs differ from conventional batteries and hydrogen-based fuel cells in that they take a novel approach to energy storage and usage. Their distinct benefits and ability to completely alter the energy landscape are shown through a comparative analysis. Unlike hydrogen-based fuel cells, DLFCs do not require fuel reforming, simplifying and expediting the energy conversion process. This correlates to increased efficiency because DLFCs directly use liquid fuels like methanol and ethanol, offering a simpler and more effective route for converting energy. DLFCs differ from regular batteries in that they have a higher energy density and a longer runtime [7]. This benefit is especially important when it comes to portable devices because users want longer gadget usage and fewer recharges. As DLFCs may use fuels with higher energy densities, their application potential is further increased, making them a strong replacement for both batteries and conventional fuel cells. Certainly, the following Table 8.1 lists the benefits and potential effects of several DLFC designs:

As they are designed for certain fuel sources and uses, several DLFC models each have special advantages. Methanol DLFCs are ideal for backup power and on-the-go electronics because they allow quick recharging and increased runtime for portable devices. Due to ethanol's renewable status,

TABLE 8.1

Lists the Benefits and Potential Effects of Several Direct Liquid Fuel Cell (DLFC)

DLFC Type	Advantages and Impact
Methanol	High energy density allows for extended runtime in portable devices; rapid refuelling enhances convenience; potential for clean backup power systems.
Ethanol	Similar advantages to methanol DLFCs; ethanol is renewable and more environmentally friendly as a fuel.
Hydrocarbon	Exceptionally high energy density enables longer operation and reduced recharging; suitable for specific transportation and off-grid applications.
Application-Specific	Customized DLFC designs for specific industries offer tailored advantages, such as greater durability for rugged use or optimized energy conversion for electronics.

ethanol DLFCs have similar advantages while being more environmentally benign. In applications needing longer operation times, such as transportation and off-grid scenarios, hydrocarbon DLFCs perform exceptionally well in terms of energy density. Furthermore, application-specific DLFCs, which are DLFCs made specifically for a given sector, can answer that industry's needs. These specialized DLFC designs could concentrate on robustness for outdoor use or improved energy conversion for electronics, thereby increasing their influence in the targeted industries. Overall, the wide variety of DLFC types and their benefits work together to support DLFCs' transformative potential in a number of sectors and applications [8–10].

The importance of DLFCs is seen in its ability to address significant energy concerns in a variety of industries. As a result of DLFCs' use of liquid fuels, devices and systems can operate for longer periods of time while experiencing shorter refuelling times and higher energy densities. This benefit is essential in applications where dependability and sustainability are critical, such as consumer electronics and backup power systems. Additionally, DLFCs have the power to revolutionize the transportation sector by providing cleaner and more effective energy sources for unmanned aerial vehicles and electric automobiles. By enhancing mobility and reducing carbon emissions, this could influence the development of sustainable transportation in the future [11]. The transformational impact of DLFCs is supported by the direct use of liquid fuels. By streamlining the energy conversion process, this method reduces waste and increases effectiveness. As a result, by providing a link between the high energy density of liquid fuels and the advantages of fuel cells for the environment, DLFCs have the potential to completely alter the energy landscape [12]. DLFCs pave the path for a more sustainable and resilient future across industries ranging from electronics to transportation and beyond by providing a cleaner, more adaptable energy supply.

8.3 Conflicts Feature in Direct Liquid Fuel Cell

The development of DLFCs at this time points to their potential expansion in upcoming applications and their advantages over existing fuel cell technologies. However, a number of restrictions prevent the development and widespread use of DLFC systems. It is essential to quickly and effectively solve all current restrictions for DLFCs to function as a practical substitute for producing renewable energy. Batteries and fossil fuels are DLFCs' main competitors in industrial settings. This section largely focuses on identifying the common difficulties faced by various DLFC types and explores potential solutions to these problems, providing a framework for future studies.

8.3.1 Costing Catalyst and High Loading of Catalyst Necessity

The main challenge for DLFCs is the expensive cost of catalysts and their heavy reliance on high catalyst loading to ensure DLFC function. Between DLFCs and other systems like lithium-ion batteries, this twofold issue is a key differentiation. With catalyst loading percentages that are higher than those of hydrogen PEMFC systems, all DLFC variations rely on noble metal-based catalysts. To achieve a minimum power density of 23 mW cm^{-2}, the DMFC system used in commercial applications uses PtRu and Pt catalysts on the anode and cathode sides, respectively. This requires an 8 mg cm^{-2} catalyst loading on both electrodes. Scholars have looked into ways to reduce catalyst loading in an effort to solve this. Ahmad et al. [13] used just 5 mg cm^{-2} of PtRu on the anode side to produce a DMFC power density of 14 mW cm^{-2}. Similar to this, Hashim et al. [14] achieved a maximum power density of 12 mW cm^{-2} by reducing the catalyst loading to just 3 mg cm^{-2} of PtRu on the anode. Gogel et al. [15] offered a different viewpoint, noting that declining catalyst loading resulted in decreased system performance because of a decline in oxidation reaction efficiency. Pt remains crucial despite alternate catalyst proposals. Numerous researches chose high-loading catalysts, such as DDEFCs, DPFCs, DEGFCs, DGFCs, and DBFCs, to maintain operational efficacy since they recognized the significant impact of decreasing catalyst loading on DLFC performance.

The expensive noble metal Pt, which drives the elevated catalyst loading, has a considerable impact on the system's overall effectiveness. By investigating hydrogen PEMFCs with a startlingly low catalyst loading of just 1 mg cm^{-2}, Adilbish et al. [16] addressed the urgent issue of catalyst cost and loading that has not yet been answered. Furthermore, improving catalyst efficiency is still essential even with lower loading. A potential method for improving catalyst performance is the modification of the catalyst using additions like carbon nanofibers, mesoporous carbon, carbon nanotubes, fullerene, graphene, and functionalized graphene. Comparative testing of the catalytic potential of carbon nanofibers and traditional carbon black was done by Zainoodin et al. [17]. Satisfactory results were obtained, yielding 18.3 and 12.6 mW cm^{-2}, respectively, while maintaining PtRu and Pt catalyst loading at both electrode sides at 8 mg cm^{-2}. Particularly, compared to carbon black, the insertion of carbon nanofibers showed increased catalyst reaction efficiency. This choice helps to maintain outstanding DLFC performance by reducing the quantity of catalyst needed when supporting materials are added.

The efficiency of the catalyst can also be impacted by the structure of the electrode. In an active DMFC system, Almheiri et al. [18] used a PtRu catalyst with a 4 mg cm^{-2} loading to achieve a maximum power density of 104 mW cm^{-2}, where active systems theoretically perform better than passive ones. Liu et al. [19] showed improvements in DLFC performance by fabricating MEA using ultrasonic spray techniques and altering the porous shape of the catalyst. Even though the catalyst loading was the same, the MEA preparation process produced noticeable benefits. Although progress has been achieved in reducing catalyst loading and improving catalyst efficiency, further research is still required because this issue continues to be a major problem for DLFC and presents barriers to commercialization because of its close relationship to overall system costs.

8.3.2 Manufacturing Cost

In order to reduce overall system costs, research, development, and demonstration (RD&D) activities are focused on developing balance of plant (BOP) systems and fuel cell stacks that are cost-effective, as well as enhanced high-volume manufacturing techniques. Given that platinum is a significant cost component in polymer electrolyte membrane fuel cells that run on direct hydrogen, strategies focus on increasing activity and utilization, lowering the content of the current platinum group metal (PGM) and PGM-alloy catalysts, and investigating PGM-free catalyst options for long-term use [20,21]. The cost of DMFCs' architecture is a major barrier to attaining a widespread market presence across different power requirements. According to the extensive usage of expensive metal catalysts (PtRu and Pt) in quantities more than 2 mg cm^{-2} and the use of expensive Nafion-type membranes (\$600–1200 m^{-2}), DMFC production costs are higher than those of PEMFCs and Li-ion batteries. A Korean business reported in 2007 that the cost of manufacturing a 20 W DMFC for portable computers was \$333, which was ten times more expensive than making a Li-ion battery with a comparable energy capacity.

Recently, noteworthy steps have been achieved to lower costs and lengthen the operational life of DMFC systems. The Department of Energy (DOE) recently updated its technological goals for portable DMFC devices with the intention of meeting these goals by 2015. These objectives include delivering a 35% increase in system performance, 5,000 hours of operational life, and cost reductions of \$5 and \$7 per watt for systems with capacities of 100 to 250 W and 10 to 50 W, respectively, at scale. Notably, a number of businesses, like SFC and MTI, have successfully finished endurance tests lasting more than 5,000 hours for their hypothetical DMFC devices, albeit specific costs are yet unknown. Recent initiatives have mostly concentrated on cost reduction and performance improvement of individual components, with comparatively less emphasis being paid to thoroughly examining the cost advantages provided by large-scale industrial manufacturing. The economic benefits required to encourage demand acceptance could be significant if these fuel cells are produced at a scale similar to commercial production. Fuel cell stack costs are primarily influenced by expenses for the membrane, noble metal catalysts, electrodes, bipolar plates, clamping plates, current collectors, gaskets, peripherals, and assembly processes [22]. These costs take into account the financial investment, labor costs, and energy costs. It is critical to understand that the fuel cell system's modular design lends itself to efficient mass production. For manufacturers trying to attain maximum competitiveness for industrial use, output, and energy application, the cost of the DMFC system continues to be a critical challenge [23]. An exploded graphic is used to illustrate how the DMFC stack configuration developed as a result of the joint efforts of the FCH JU initiative and the European Community FP7 project DURAMET (www.duramet.eu). The following important features summarize the developments in DMFC technology made under the limited parameters of DURAMET's early development:

1. Putting in place a method for producing cathodic and anodic catalysts/electrodes that is both affordable and efficient.
2. Decrease in the number of precious metals used in the catalytic layers.
3. Release of novel membranes that are more reasonably priced and have dramatically reduced thickness and methanol permeability as compared to the industry-standard Nafion membrane.

Beyond the MEA (membrane electrode assembly), there are other aspects, however, that have a significant impact on the extensive production of the resulting DMFC stacks. Numerous factors have an impact on the overall expenses related to each individual component; therefore, careful attention is required.

8.3.3 Unproven Long-Term Durability

For systems, including DLFCs, durability is a crucial performance characteristic that necessitates careful testing in order to verify their viability for commercial use. High durability denotes superior quality systems that can operate continuously without degrading initial functionality. The prolonged response time degrades the performance of the DLFC system, causing instability brought on by catalyst agglomeration, changes in electrode porosity, and electrolyte membrane degradation [24]. Almost all DLFC systems use PtRu catalysts, which act as anode catalysts. Ru crossover to the electrolyte and cathode occurs as a result of the anode catalyst's effectiveness degrading over time as a result of catalyst degradation. As a result, oxygen reduction is impeded, decreasing the cathode process' effectiveness [25]. Degradation of the catalyst negatively impacts cathode flooding, water management, and the cathode's hydrophobic properties. Keeping performance at its peak for lengthy periods of time is crucial for fuel cell applications. The Department of Energy (DOE) has set high goals for fuel cell system lifetime under real-world operating conditions, aiming for 8,000 hours for light-duty vehicles, 30,000 hours for heavy-duty trucks, and 80,000 hours for distributed power systems. Ensuring system dependability and resilience in the face of dynamic and challenging operational settings is essential, particularly in demanding applications. Practical scenarios include elements like start-stop cycles, freeze-thaw cycles, fuel and air contaminants, humidity variations, and dynamic load fluctuations. The chemical and mechanical stability of the materials and components used in fuel cell systems is put under pressure by these circumstances. The focus of research, development, and demonstration (RD&D) activities is on identifying and understanding the mechanisms of fuel cell degradation, as well as developing materials and methods to lessen their effects, including addressing problems like the movement of oxygen towards the cathode layer.

This problem leads to a gradual decline in the system's performance. In a testing operation by Cheng et al. [19] that lasted 200 hours, cell performance indicated a decline; after 1,002 hours, total operation showed a noticeable decline. The prolonged operation causes an increase in the catalyst's mean particle size within each electrode as PtRu particles amass. Anode electrode surface saturation with metal oxides after continuous operation reduced the catalyst's specific surface area, which, in turn, decreased fuel cell activity, as shown in separate research. After a certain amount of processing time, according to Schoekel et al. [26] verified Ru crossover from the anode to the cathode (Figure 8.2). A general deterioration in DMFC performance results from the translocated Ru's interference with the cathode's response.

After 50 test cycles with Pd-nanorods acting as the anode catalyst, Cerritos et al. noticed a declining performance trend in DEFCs, which caused the power density to drop from 6 mW cm^{-2} to under 5 mW cm^{-2}. Long-term use caused the catalyst's structural makeup to shift, which decreased performance. According to Zhang et al. [28] the DGFC system experienced performance degradation after 100 test cycles, with a drop in the final power density of more than 50% from the initial measurement (Figure 8.3). As emphasized in Xing et al.'s [29] work, the DDEFC also saw a decline in performance during prolonged operation. The cell voltage and power density both dropped over time, from 0.59 V to 0.52 V and 37.4 mW cm^{-2} to 12.0 mW cm^{-2}, respectively. After 200 hours of continuous use, Filanovsky [30] stated that the DHFC system had completely failed to function. Bubble accumulation

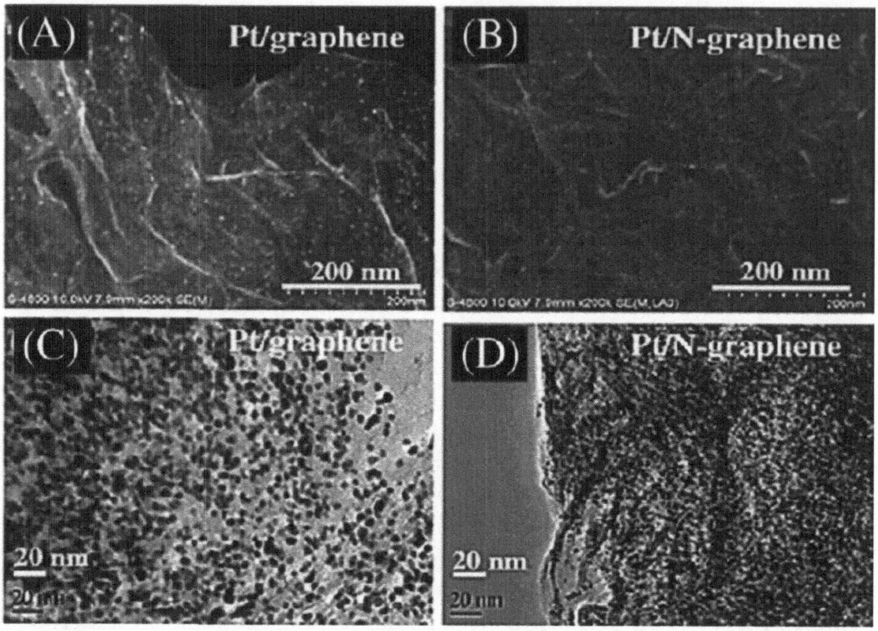

FIGURE 8.2
Pt/graphene and Pt/N-graphene represented in (a and b) SEM images and TEM images (c and d) [27].

was brought on by structural degradation and decreased MEA adhesion. However, performance started to deteriorate after that. Subsequent design improvements increased operation to 1,000 hours. As a result, almost all DLFC systems run into issues and degrade their performance over time. Future research should concentrate on attaining better results, especially in stability and durability performance evaluations.

According to research by Kimiaie et al. [31], the DMFC system V3.3–2 offers a potential path for the widespread use of DMFC devices because it demonstrated a robust endurance of 20,000 lifetime cycles after extensive life testing with various loading patterns. The hybrid DMFC system V3.3–2 is the result of this innovation, which combines a 1.0 kW DMFC system with a 45 Ah high-power lithium-ion battery pack. This integrated setup, which has a 7 kW total output capacity, can replace a Class 3 truck. This energy generation system has positive aspects over traditional lead-acid batteries, such as greater capacity (allowing 24 hours of use from a 20 L methanol cartridge as opposed to eight hours of battery capacity) and improved accessibility (replacing methanol cartridges only takes a few minutes rather than the prolonged charging hours required for batteries). The DMFC stack must have a minimum lifespan of 10,000 hours nevertheless in order to guarantee the DMFC system V3.3–2's economic sustainability. In the first 25,600 hours

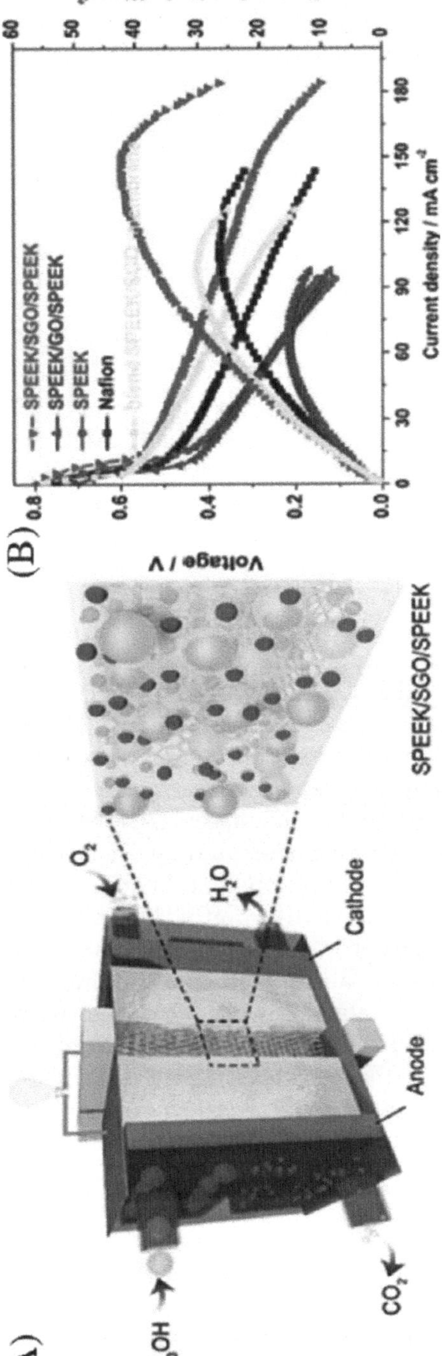

FIGURE 8.3
(A) Schedule of SPEEK/RGO/SPEEK membrane and semi-passive DMFC. (B) The DMFC performing with the various polymer electrolyte membranes.

of the worldwide assessment, the integrated system has been running effectively for 20,000 hours.

The 20,000-hour period fits with the normal life of vehicles used in the container handling industry. Its appropriateness for use in large trucks is demonstrated by the tested and built DMFC system V3.3–2, which goes above and beyond the criteria of market commercialization. The passive single-cell DMFC model, a six-cell direct methanol stack that is passive and air-breathing, serves as the model for the creation, assembly, and evaluation of the six-cell DMFC stack. This was done in Masdar et al.'s [32] study. It substantially eliminates the requirement for liquid methanol replenishment during stack development since a hexagonal construction allows for significant methanol storage, extending DMFC operational duration without intervention. The DMFC stack is revealed, which can output 500 mW at 1.5 V while only requiring a 5 M methanol solution to operate at ambient temperature. Examining the MEA structure reveals that the stack has been continuously active for more than 3,000 hours. The addition of a sizable storage tank within the stack increases operational duration to over 40 hours without the need for methanol introduction or replenishment, but this is in contrast to the three to four hours of service possible with a single cell, leading to comparable power production declines of about 25%. The short lifespan of the DMFC system, according to Muller et al., is the system's main problem.

DMFC systems significantly degraded in less than 1,000 hours, whereas individual cells showed the ability to function for long periods of time. A gradual evolution over six generations of DMFC systems in the kilowatt power range led to the identification of key degradation drivers. In order to understand the fundamental causes of progressive degradation over several hundred or thousand hours, it was first necessary to eliminate quick degradation triggers [33]. The need for understanding the linkages between cells and the control system also became apparent. In order to ensure a favorable environment for all cells, this led to the careful observation of optimal operational algorithms and repeatable production procedures. This prevented a higher anode potential from causing ruthenium corrosion and ultimately contaminating the cathode catalyst. It was crucial to make sure that the peripheral system parts and the stack both had excellent corrosion resistance and were free of impurities that could permeate the membranes [34]. Finally, a DMFC architecture was put through 20,000 hours of cumulative practical load testing, specifically using a pallet jack.

8.3.4 Hydrogen Evolution

Research and development (R&D) efforts are aimed at improving the efficiency and performance of fuel cells through the use of new materials and integration technologies. This includes multiple aspects, such as the development of ion-exchange membrane electrolytes with improved efficiency, durability, and cost-effectiveness; the enhancement of membrane electrode assemblies (MEAs) to achieve high power density through the integration of cutting-edge MEA components; leveraging modeling techniques to comprehend system design and operational conditions; and creating stacks with high efficiency at rated power alongside high-penetration. Hydrogen evolution is a key issue in the development of the direct borohydride fuel cell (DBFC) stack and its commercialization [35]. The main cause of stacking loss in DBFCs is hydrogen evolution, which results in an unequal distribution of fuel among the cells in the stack. Higher initial borohydride concentrations cause accelerated hydrogen evolution, which has a negative effect on DBFC performance and fuel efficiency. A lot of work has been put on reducing borohydride hydrolysis, which is the main pathway for hydrogen evolution. Incorporating substances that have inhibitory effects on hydrogen evolution into the BH4 solution and finding anode catalysts that enable complete electro-oxidation of BH4 without inducing hydrolysis reactions are two methods for getting around this restriction [36]. In addition, to effectively address these issues, a focus on flow field design optimization and improving anode structure, such as the creation of hydrophobic holes, is still essential.

Gao et al. [37] and Chen et al. [38] used nickel-based hydrogen storage alloys as anode materials and added Pd/C catalyst to the FMH-Ni anode in an effort to prevent hydrogen evolution. Martins et al. [39] suggested using tetraethylammonium hydroxide (TEAH) and thiourea (TU) as inhibitors for borohydride hydrolysis processes. Thiourea (TU) was included by Celik et al. [40] as an additive in sodium borohydride solutions to reduce anodic hydrogen evolution on Pd, which improved the efficiency of direct borohydride fuel cells. By reducing the impact of hydrogen evolution by a well-designed anode shape, Wang et al. [41] sought to improve stack performance.

8.4 Chemical Safety and Environmental Concern of the Fuel

Numerous different fuel types can be used with the DLFC system; however, not all of them are suitable owing to safety issues. Because of the increased toxicity of some fuels, safety becomes a major challenge [42]. One very dangerous fuel for human exposure is hydrazine, which is widely acknowledged as one of the most dangerous substances for use in engines. It has been recorded that exposure to hydrazine for six months can be lethal, and it can also cause harm to the kidneys, lungs, and neurological system in those exposed to it. Therefore, a major barrier to the commercialization of DHFCs is still hydrazine's toxicity. Particularly, fuels containing carbon have a propensity to emit CO_2, and there are other fuels with low levels of toxicity that are within the realm of chemical safety regulations, such as methanol and dimethyl ether. However, there are worries about the corrosive characteristics of formic acid, raising safety issues that limit its commercial viability [43].

The corrosive nature necessitates careful handling to avoid skin or eye contact, as well as attentive storage system maintenance to reduce the possibility of user exposure. Additionally, fuels with flammable characteristics, such alcohol-based fuels, cause safety concerns that call for aggressive mitigation.

Utilizing combustible fuels requires careful handling to reduce the risk of fire. To guarantee the secure commercialization of such fuels, it is crucial to build a fire-resistant system for users. Concerns are raised by toxic byproducts, such as carbon monoxide produced by fuel reactions in DMFC, DEFC, and DFAFC systems. These substances may have a negative effect on the catalyst anode, leading to decreased system durability and activity. In order to stop catalyst degradation, it is crucial to remove hazardous byproducts from the system. A lot of work has been put into creating catalysts that don't interact with these byproducts. In addition, other harmful byproducts, such as formaldehyde produced by DDEFC, offer serious hazards and may result in blindness or damage to the visual nerve system [44,45]. As a byproduct, DEGFC produces glycolic acid, whereas DHFC produces ammonia, both of which raise questions about contamination of the air and water. Treatment operations must be carried out to make these byproducts harmless before releasing them into the environment. However, these mitigation measures come at an additional expense for residual treatment, raising the already high prices of DLFC. Therefore, it is necessary to make deliberate efforts to either completely eradicate or drastically lessen these negative side effects.

The most significant possible risks related to liquid fuels are included in Table 8.2, with all information presented taken from official sources provided by chemical corporations, specifically "material safety data sheets." Table 8.2 initially shows that all liquid fuel components have negative qualities, but it is possible to distinguish between these negative qualities.

Because of its numerous negative effects—including its carcinogenicity, severe toxicity, and considerable negative effects on society and the environment—hydrazine usage is strongly prohibited. Due to its flammable and polluting properties, hydrazine is not recommended for frequent consumer usage because it is both unpleasant and dangerous. The one and only significant flaw in dimethyl ether is that it is highly flammable. The less harmful alternative is borohydride, while being poisonous even in liquid form. As sodium hydroxide is present, the borohydrate solution is a very concentrated aqueous alkaline solution with a pH that exceeds 13. As a substance, sodium hydroxide is noncombustible, moderately corrosive, and irritant to the skin. Given the hazardous qualities listed in Table 8.2 and the fact that each fuel component contains hazards, choosing the "ideal" liquid fuel is a difficult procedure. The best fuel to use becomes an issue of balancing negative effects with fuel cell performance. It is categorically advised against using hydrazine in light of this judgment. In conclusion, because of its simple manufacturing procedure, ethanol has won experts' trust as a promising fuel source for small-scale fuel cells. A preferable option, ethanol has somewhat less harmful characteristics while still being energy-efficient.

TABLE 8.2

Tabulated List of Significant Risks Related to Different Types of Liquid Fuels, Categorized by Storage, Transportation, Usage, and Regulatory Practices

Liquid Fuel Type	Storage Risks	Transportation Risks	Usage Risks	Regulatory Risks
Gasoline	Flammability, vapor release, leaks	Accidents during transport, spillage, emissions	Fire hazards, toxic exposure, emissions	Regulatory compliance, environmental impact
Diesel	Combustibility, fuel degradation	Spills, accidents, emissions	Health effects from exhaust, fuel quality concerns	Emissions standards, fuel quality regulations
Ethanol	Fire hazards, water absorption, phase separation	Leaks, corrosive effects	Vehicle compatibility, health effects	Ethanol content regulations, fuel quality standards
Methanol	Flammability, toxicity, leaks	Vapor release, accidents	Toxic exposure, emissions, engine compatibility	Regulatory limits on methanol content
Hydrogen Peroxide	Decomposition hazards, corrosivity, contamination	Spill hazards, instability during transport	Corrosion, chemical hazards, safety precautions	Regulations on handling and transport of chemicals
Biodiesel	Oxidation, microbial growth, viscosity changes	Spills, contamination, temperature sensitivity	Engine compatibility, emissions, cold flow properties	Biodiesel blend standards, emissions regulations
Jet Fuel	Contaminant growth, degradation, freezing point issues	Spills, transportation incidents	Combustibility, emissions, aviation safety	Aviation fuel quality standards, safety regulations
Kerosene	Flammability, toxicity, exposure risks	Spills, transportation hazards	Combustibility, inhalation risks, emissions	Regulatory standards for storage and handling
Propane	Flammability, asphyxiation risks	Leaks during transport, handling hazards	Fire hazards, inhalation risks, appliance safety	Propane storage regulations, safety standards
Natural Gasoline	Flammability, vapor release, contamination	Accidents during transport, leakage hazards	Air pollution, health effects, combustion risks	Emissions regulations, fuel quality standards

8.5 Applications of Direct Liquid Fuel Cells

Researchers and academics are interested in DLFCs as a potential replacement for lithium-ion batteries and traditional fossil fuel–based systems. This tendency is a result of lithium-ion batteries' shortcomings, which include their inability to produce significant power and sustain prolonged operation due to insufficient energy density and their limited runtime capacities. Systems that rely on fossil fuels, however, have a higher energy density but also cause environmental damage. The development of DLFCs as an appealing response to these complex problems has attracted a great deal of attention from the commercial and research worlds. The DMFC and DEFC stand out as the most advanced and well-established DLFC variations, with additional iterations still in the early stages of development. Toshiba has made a huge step towards commercializing DMFC by unveiling a ground-breaking prototype designed specifically for laptops. This innovation demonstrates the potential of DLFCs to effectively power contemporary electronics while limiting their environmental impact. It can generate 12 W of power, run continuously for five hours, and weigh under 1 kg.

Smartphones, laptops, tablets, and wearable technology are just a few of the portable electronics that can now be powered by DLFCs (Figure 8.4). DLFCs are a tempting alternative to conventional batteries for mobile charging because they provide higher energy densities and longer operational times. To prove their superiority over rival technologies, particularly conventional batteries, a wide range of enterprises and academic groups are actively working to develop various portable fuel cell technologies. Applications for DLFCs can be found in consumer electronics, computers, battery chargers, external power sources, and even military settings. They are particularly important for devices requiring high power density and extended usage intervals. It is noteworthy that Lilliputian Solutions obtained $5 million in finance to enable the equipment procurement for its Wilmington factory specializing in USB charging solutions. Its importance is further highlighted by Intel Capital's equity investment in Lilliputian.

By starting shipments of miniature hydrogen fuel cells and associated filling stations for model hobby radio-controlled automobiles, Horizon Fuel Cell Technologies accomplished a significant milestone. The company's commitment to innovation included the development of a portable fuel cell battery charger for the consumer electronics market. Panasonic strengthened its relationship with SFC Energy by approving the Toughbook models for smooth fuel cell integration. Plans for a hybrid fuel cell technology from Neah Power Systems were presented with the intention of charging lithium-ion batteries found in consumer electronics. For a complete list of international companies that specialize in portable fuel cells, see Table 8.3. A list of multinational businesses that specialize in portable fuel cells is included in this table, together with information on their main areas of emphasis and

FIGURE 8.4
Portable electronics that be powered by DLFCs.

standout products. This list is not all-inclusive and includes a variety of businesses active in various facets of the portable fuel cell market.

The military and defense industries have great potential for DLFCs, which may be used in lightweight and effective energy sources for defense systems, such as unmanned aerial vehicles (UAVs) and portable power packs for soldiers. In addition, by supplying electric vehicles (EVs) with power and expanding their range, DLFCs have potential for the automotive sector. Their benefit over conventional batteries is quick refilling, which makes them appropriate for long-distance travel. DLFCs are advantageous for remote power generation because they provide dependable energy options for off-grid or rural areas without access to conventional electricity sources. They can be used in systems like remote weather stations, monitoring equipment,

TABLE 8.3

List of International Companies that Specialize in Portable Fuel Cells

Company Name	Country	Focus Areas	Notable Products
SFC Energy AG	Germany	Military, Industry, Consumer Electronics	EFOY Pro Series, JENNY Fuel Cell
Ballard Power Systems	Canada	Transportation, Backup Power	FCgen®-LCS Fuel Cell Stack
Toshiba	Japan	Consumer Electronics, Backup Power	Dynario™ Direct Methanol Fuel Cell Charger
Horizon Fuel Cell Technologies	Singapore	Consumer Electronics, Education, Industry	Hydrostik Pro, H-500 Fuel Cell Stack
Neah Power Systems	USA	Consumer Electronics, Military, Medical	BuzzBar™, PowerChip® Fuel Cell
Ultracell Corporation	USA	Military, Medical, Backup Power	XX25™, XX55™ Fuel Cell
PolyFuel	USA	Consumer Electronics, Military, Industry	DMFC MEAs, Fuel Cell System Components
Intelligent Energy	UK	Consumer Electronics, Transportation	Upp® Fuel Cell Portable Charger, Gen4 Fuel Cell
myFC AB	Sweden	Consumer Electronics, Backup Power	JAQ Hybrid Fuel Cell Charger
Antig Technology Co., Ltd.	China	Consumer Electronics, Portable Chargers	JAQ Hybrid Fuel Cell Charger

and communication networks. DLFCs are essential components of backup power systems because they provide reliable, environmentally friendly emergency power for crucial buildings, including data centres, hospitals, and communication hubs. Conventional diesel generators, which frequently emit noise and dangerous fumes, are replaced by this environmentally aware option. Indeed, the following Table 8.4 provides a thorough overview of the uses for DLFCs. This table gives a thorough explanation of the numerous areas where DLFCs are having a big impact, as well as the advantages and effects they have on each industry.

Applications for DLFCs are numerous and provide revolutionary advantages. They supply electricity to portable gadgets, extending device life and allowing quick refilling. They guarantee dependable, clean energy in backup power systems in times of need. In addition to renewable energy sources, DLFCs help generate energy off the grid in rural areas. Sustainable EV power and longer UAV flight periods are advantageous for the transportation sector. They improve reliability and require less maintenance, empowering tiny gadgets and medical implants [46]. DLFCs support consumer electronics, improve communication systems, enable remote monitoring, and support defense operations. Their potential to transform numerous sectors while promoting efficiency and sustainability is demonstrated by their capacity to increase automotive range and enhance healthcare.

TABLE 8.4

Overview of the Uses for Direct Liquid Fuel Cells (DLFCs)

Application	Description	Benefits and Impact
Portable Electronics	DLFCs offer extended runtime and rapid refuelling for devices like smartphones, laptops, tablets, and wearables.	Longer device usage, reduced recharging frequency, enhanced mobility.
Backup Power Systems	DLFCs provide reliable and clean backup power for critical infrastructure during outages and emergencies.	Enhanced resilience of essential services, reduced service disruptions.
Distributed Energy Generation	DLFCs serve as efficient, off-grid power sources, complementing intermittent renewables like solar and wind.	Reliable energy supply in remote areas, establishment of energy hubs.
Transportation	DLFCs power electric vehicles (EVs) and unmanned aerial vehicles (UAVs), offering quick refuelling and extended range.	Sustainable mobility, longer flight times for UAVs, reduced reliance on fossil fuels.
Micro and Miniature Devices	DLFCs provide compact and reliable power for sensors, medical implants, and microrobots in remote or inaccessible areas.	Extended device lifespan, reduced maintenance needs, enhanced functionality.
Military and Defense	DLFCs offer lightweight, efficient power solutions for soldiers, UAVs, and defense systems in challenging environments.	Enhanced mission capabilities, reduced logistical burden, quieter and cleaner operations.
Consumer Electronics	DLFCs power gadgets like smartwatches, headphones, and handheld gaming devices, extending usage and convenience.	Longer device operation, reduced need for frequent charging, enhanced user experience.
Remote Weather Stations	DLFCs provide reliable energy for distant weather monitoring stations, ensuring continuous data collection.	Uninterrupted data gathering, enhanced weather forecasting accuracy.
Telecommunications	DLFCs offer clean backup power for communication systems, preventing service disruptions during power outages.	Continuous communication services, reduced downtime and impact on emergency response.
Automotive Range Extenders	DLFCs enhance the range of electric vehicles by providing quick refuelling and extended driving capabilities.	Increased EV range, reduced "range anxiety," improved convenience for long-distance travel.
Healthcare and Medical Devices	DLFCs power medical implants, enabling long-lasting and reliable operation for patient monitoring and treatment.	Extended device lifespan, reduced need for surgeries or replacements, improved patient care.

8.6 Current Achievements and Ongoing Research

The development of DLFCs has been fueled by major accomplishments and continuing research initiatives. These accomplishments demonstrate the substantial advancements made in this ground-breaking energy technology, and current research demonstrates the continual exploration and improvement of its possibilities [47,48]. We obtain a thorough grasp of the development and potential of DLFCs as a cutting-edge solution in the field of energy generation and storage by looking into these successes and research endeavors.

8.6.1 Successes in DLFC Development and Implementation

In numerous fields, the development of DLFC technology has resulted in outstanding accomplishments. Researchers and engineers have worked diligently to develop practical proof-of-concept DLFC systems and working prototypes. They have unlocked innovations that advance the technology by tackling complex technical challenges like improving electrode materials, improving fuel delivery systems, and improving cell efficiency. These successes have made it possible for DLFCs to be used in the real world, marking an important step towards their widespread commercialization. Particularly interesting is how well DLFCs have been incorporated into real-world uses. Case studies illustrate situations where DLFCs have successfully adapted to their surroundings, such as when powering consumer devices or acting as dependable backup power supplies. The versatility and practical practicality of DLFCs are illustrated by these real-world examples, highlighting their potential to fundamentally alter how we power our infrastructure and devices. These accomplishments highlight the real advancements made and the encouraging trajectory towards a future that is more energy-efficient and sustainable as DLFC technology continues to advance.

8.6.2 Notable Use Cases and Industry Applications

The variety of uses for which DLFCs have potential is both fascinating and significant. With advantages over conventional batteries, including longer runtime and quick recharging, DLFCs have made a significant impact in the field of powering portable electronic gadgets. Additionally, they have demonstrated their worth as crucial elements of backup power systems, enhancing the resilience of vital infrastructure and providing sustainable energy options in times of crisis. Additionally, investigating DLFCs for distributed energy generation in off-grid and distant locations offers enormous promise. In such circumstances, their special capacity to extract energy from liquid fuels proves to be very useful. Additionally, DLFCs are gaining popularity in the transportation industry, displaying their potential to power UAVs and electric cars, opening the door to more sustainable mobility and longer flying times. The adaptability and versatility that DLFCs bring to a variety of industries are best demonstrated by these numerous use cases, which present creative ways to address various energy needs.

8.6.3 Collaborative Efforts in Advancing DLFC Technology

Collaborations between research institutes, industry, and academia have greatly advanced DLFC technology. Collaborations have sped up the development of DLFCs by facilitating the flow of information, skills, and resources. The groundwork for comprehending the fundamentals of DLFC has been built by academic research, and this knowledge has been transformed into useful applications through industry partnerships. The funding of research, the development of standards, and the execution of larger-scale trials have all been made possible through collaborative endeavors involving governmental organizations, private businesses, and academic institutions. Research on DLFC has been multidisciplinary, resulting in advances in electrochemistry, materials science, and fuel cell engineering. The mutually beneficial interaction between these industries keeps DLFC innovation moving ahead. This section emphasizes the hopeful future of direct liquid fuel cells by focusing on the accomplishments made in DLFC development and the continuing research initiatives. These successes highlight how DLFCs have the ability to transform energy storage and use as well as solve critical problems across numerous industries.

8.7 Future Targets for Application Direct Liquid Fuel Cells

In order to promote the commercialization of portable fuel cells, the US Department of Energy has established precise goals for a number of metrics, including price, toughness, power, energy, and mean time between failures. The production quantities must meet these goals, which are listed in Table 8.5. In order to promote the commercialization of portable fuel cells, the US Department of Energy has set clear goals for a number of indicators. These objectives involve topics like cost, durability, power density, energy density, and mean time between failures and vary depending on the power output of the fuel cell systems. For portable fuel cells to be more widely adopted and integrated into many applications, it is imperative that these goals are met.

According to calculations using a lower heating value (LHV) of 638.1 kJ mol^{-1}, liquid methanol has an energy density of 5.53 Wh g^{-1} and 4.35 Wh cm^{-3}. The effective energy densities of methanol are computed to be 1600 Wh kg^{-1} and 1300 Wh L^{-1}, which surpass the 2015 targets by 2.5 and 1.4 times,

TABLE 8.5

Objectives Set for the Commercialization of Portable Fuel Cells

Metric	Target for < 2 W Systems	Target for 10–50 W Systems	Target for 100–250 W Systems
Cost (USD/kWh)	\$50	\$30	\$20
Durability (hours)	1,000	5,000	10,000
Power Density (W/cm^{-2})	2	2	2
Energy Density (Wh/L)	300	300	300
Mean Time Between Failures (hours)	1,000	2,000	5,000

respectively, assuming a DMFC electrical efficiency of 30%. Carrying more liquid methanol will make it easier to achieve these goals because it has a big impact on the amount of energy and energy density targets that can be attained. Targets relating to specific power and power density, on the other hand, are more crucial and challenging.

DLFCs applications in the future show great promise for resolving issues with existing fuel cell technology and conventional energy storage systems. The ability of DLFCs to use liquid fuels directly without the need for hydrogen reforming has many benefits, including improved energy density, simpler storage, and potential cost effectiveness. By satisfying the growing need for increased battery life in products like smartphones and laptops, DLFCs could completely change the landscape in the field of portable electronics. The user experience and mobility of a device can be improved with DLFCs, which provide longer runtimes and speedier refilling times. With the ability to deliver reliable and clean energy for essential infrastructure and emergency circumstances, DLFCs are prepared to solve critical needs in backup power systems. With this potential resilience improvement, service interruptions might be reduced, and uninterrupted operations might be provided in the event of a power outage.

The additional path for DLFCs to take is distributed energy generation, especially in off-the-grid or rural locations. By providing dependable and effective electricity, DLFCs could support intermittent renewable energy systems, promoting sustainable energy solutions. Because DLFCs use liquid fuels and can recharge more quickly than other fuels, the transportation industry could benefit from their use in tiny electric vehicles or as range extenders for electric cars. Additionally, DLFCs have the potential to change drone and UAV capabilities, enabling longer flight periods and quicker recharging, revolutionizing industries like delivery and surveillance. The benefits of DLFCs extend to small and micro devices, offering a reliable and long-lasting power source for sensors, implants, and microrobots. These devices have new possibilities across industries thanks to their small size and reliance on liquid fuels, which also reduces the need for maintenance. Despite being hopeful, overcoming obstacles, such as improving efficiency, durability, and cost effectiveness, is necessary for the full fulfillment of DLFCs' promise. The widespread deployment of DLFCs will be accelerated by ongoing research and collaboration between academia and industry.

DLFCs have become a revolutionary force that has the potential to transform industries that depend on energy sources. They show great promise for a wide range of applications thanks to their unique capacity to directly extract energy from liquid fuels. While overcoming technical obstacles continues to be an important milestone, the development of DLFC technology has the potential to fundamentally alter how we store and use energy. DLFCs have the potential to significantly alter sectors like portable electronics, transportation, and distributed energy generation by facilitating more effective and sustainable energy use. The global quest for cleaner, more reliable energy systems is consistent with this disruptive potential. As scientists and inventors struggle to advance DLFC technology, we get closer to a day when energy use is more environmentally conscious, efficient, and responsible, advancing us towards a better ability to cohabit with our environment and make the best use of our resources.

8.8 Conclusion

The potential for DLFCs to replace traditional batteries and hydrogen-based fuel cells in a variety of industries is enormous. They have a variety of uses because of their unique ability to use liquid fuels like methanol and ethanol without the requirement for fuel reformation. DLFCs could transform power sources for portable electronics, enabling longer battery life and easier refilling, while backup power systems could have access to reliable, clean energy. Due to their quick recharging and longer flight periods, DLFCs may find use in small electric cars and unmanned aerial vehicles (UAVs) for transportation. They may also improve remote, or off-grid, installations for distributed energy generation. The promise of DLFCs also includes implantable technology and micro devices. Despite persistent issues with costs and efficiency, industry cooperation and continued research can help them reach their full potential. In the end, DLFCs show potential as adaptable and revolutionary energy solutions, altering our usage of energy for a sustainable future.

References

[1] N. Sazali, W. N. W. Salleh, A. S. Jamaludin, and M. N. M. Razali, "New perspectives on fuel cell technology: A brief review," *Membranes*, vol. 10, no. 5, p. 99, May 2020.

[2] U. B. Demirci, "Direct liquid-feed fuel cells: Thermodynamic and environmental concerns," *J. Power Sources*, vol. 169, no. 2, pp. 239–246, Jun. 2007.
[3] B.C. Ong, S. K. Kamarudin, and S. Basri, "Direct liquid fuel cells: A review," *Int. J. Hydrogen Energy*, vol. 42, no. 15, pp. 10142–10157, 2017.
[4] J. Y. Lee, R. Verayiah, K. H. Ong, A. K. Ramasamy, and M. B. Marsadek, "Distributed generation: A review on current energy status, grid-interconnected PQ issues, and implementation constraints of DG in Malaysia," *Energies*, vol. 13, no. 24, p. 6479, Dec. 2020.
[5] L. Fan, Z. Tu, and S. H. Chan, "Recent development of hydrogen and fuel cell technologies: A review," *Energy Rep.*, vol. 7, pp. 8421–8446, Nov. 2021.
[6] I. Petreanu et al., *Fuel Cells: Alternative Energy Sources for Stationary, Mobile and Automotive Applications*. IntechOpen, 2020.
[7] N. Sazali, W. N. W. Salleh, A. S. Jamaludin, and M. N. M. Razali, "New perspectives on fuel cell technology: A brief review," *Membranes*, vol. 10, no. 5, p. 99, May 2020.
[8] K. K. Maniam, R. Chetty, R. Thimmappa, and S. Paul, "Progress in the development of electrodeposited catalysts for direct liquid fuel cell applications," *Appl. Sci.*, vol. 12, no. 1, p. 501, Jan. 2022.
[9] J. A. Gómez and D. M. F. Santos, "The status of on-board hydrogen storage in fuel cell electric vehicles," *Designs*, vol. 7, no. 4, p. 97, Aug. 2023.
[10] Y. Manoharan et al., "Hydrogen fuel cell vehicles; Current status and future prospect," *Appl. Sci.*, vol. 9, no. 11, p. 2296, Jun. 2019.
[11] V. Velisala, G. N. Srinivasulu, B. S. Reddy, and K. V. K. Rao, "Review on challenges of direct liquid fuel cells for portable application," *World J. Eng.*, vol. 12, no. 6, pp. 591–605, Dec. 2015.
[12] R. Oliveira, J. Santander, and R. Rego, "Overview of direct liquid oxidation fuel cells and its application as micro-fuel cells," *Adv. Electrocatal. Low-Temperature Fuel Cells*, pp. 129–174, 2018.
[13] M. M. Ahmad, S. K. Kamarudin, W. R. W. Daud, and Z. Yaakub, "High power passive μDMFC with low catalyst loading for small power generation," *Energy Convers. Manag.*, vol. 51, no. 4, pp. 821–825, Apr. 2010.
[14] N. Hashim, S. K. Kamarudin, and W. R. W. Daud, "Design, fabrication and testing of a PMMA-based passive single-cell and a multi-cell stack micro-DMFC," *Int. J. Hydrogen Energy*, vol. 34, no. 19, pp. 8263–8269, Oct. 2009.
[15] V. Gogel, T. Frey, Z. Yongsheng, K. A. Friedrich, L. Jörissen, and J. Garche, "Performance and methanol permeation of direct methanol fuel cells: Dependence on operating conditions and on electrode structure," *J. Power Sources*, vol. 127, no. 1–2, pp. 172–180, Mar. 2004.
[16] G. Adilbish, J. W. Lee, Y. S. Jang, H. G. Lee, and Y. T. Yu, "Preparation of Pt/C electrode with double catalyst layers by electrophoresis deposition method for PEMFC," *Int. J. Hydrogen Energy*, vol. 39, no. 7, pp. 3381–3386, Feb. 2014.
[17] A. M. Zainoodin, S. K. Kamarudin, and W. R. W. W. Daud, "Electrode in direct methanol fuel cells," *Int. J. Hydrogen Energy*, vol. 35, no. 10, pp. 4606–4621, May 2010.
[18] S. Almheiri and H. Liu, "Separate measurement of current density under land and channel in Direct Methanol Fuel Cells," *J. Power Sources*, vol. 246, pp. 899–905, Jan. 2014.
[19] G. Liu, M. Wang, Y. Wang, Z. Tian, and X. Wang, "A novel anode catalyst layer with multilayer and pore structure for improving the performance of a direct methanol fuel cell," *Int. J. Energy Res.*, vol. 37, no. 11, pp. 1313–1317, Sep. 2013.

[20] C. S. Gittleman, A. Kongkanand, D. Masten, and W. Gu, "Materials research and development focus areas for low cost automotive proton-exchange membrane fuel cells," *Curr. Opin. Electrochem.*, vol. 18, pp. 81–89, Dec. 2019.
[21] S. T. Thompson et al., "Direct hydrogen fuel cell electric vehicle cost analysis: System and high-volume manufacturing description, validation, and outlook," *J. Power Sources*, vol. 399, pp. 304–313, Sep. 2018.
[22] M. Javaid, A. Haleem, R. P. Singh, and R. Suman, "Substantial capabilities of robotics in enhancing industry 4.0 implementation," *Cogn. Robot.*, vol. 1, pp. 58–75, Jan. 2021.
[23] T. Ahmad et al ., "Energetics systems and artificial intelligence: Applications of industry 4.0," *Energy Rep.*, vol. 8, pp. 334–361, Nov. 2022.
[24] J. A. Gómez, D. M. F. Santos, R. Salkuti, D. Zhu, J. A. Gómez, and D. M. F. Santos, "The status of on-board hydrogen storage in fuel cell electric vehicles," *Designs*, vol. 7, no. 4, p. 97, Aug. 2023.
[25] M. Mansor, S. N. Timmiati, K. L. Lim, W. Y. Wong, S. K. Kamarudin, and N. H. Nazirah Kamarudin, "Recent progress of anode catalysts and their support materials for methanol electrooxidation reaction," *Int. J. Hydrogen Energy*, vol. 44, no. 29, pp. 14744–14769, Jun. 2019.
[26] A. Schoekel, J. Melke, M. Bruns, K. Wippermann, F. Kuppler, and C. Roth, "Quantitative study of ruthenium cross-over in direct methanol fuel cells during early operation hours," *J. Power Sources*, vol. 301, pp. 210–218, Jan. 2016.
[27] Y. Zhao et al., "Facile single-step preparation of Pt/N-graphene catalysts with improved methanol electrooxidation activity," *J. Solid State Electrochem.*, vol. 17, no. 4, pp. 1089–1098, Apr. 2013.
[28] Z. Zhang, L. Xin, and W. Li, "Supported gold nanoparticles as anode catalyst for anion-exchange membrane-direct glycerol fuel cell (AEM-DGFC)," *Int. J. Hydrogen Energy*, vol. 37, no. 11, pp. 9393–9401, Jun. 2012.
[29] L. H. Xing, G. P. Yin, Z. B. Wang, S. Zhang, Y. Z. Gao, and C. Y. Du, "Investigation on the durability of direct dimethyl ether fuel cell. Part I: Anode degradation," *J. Power Sources*, vol. 198, pp. 170–175, Jan. 2012.
[30] B. K. Filanovsky, "Liquid fuel cells," *NCBI—Natl. Institutes Heal. J Power Sources*, vol. 246, no. 9, pp. 423–429, 2014.
[31] N. Kimiaie et al ., "Results of a 20 000 h lifetime test of a 7 kW direct methanol fuel cell (DMFC) hybrid system—degradation of the DMFC stack and the energy storage," *Energy Environ. Sci.*, vol. 7, no. 9, pp. 3013–3025, Aug. 2014.
[32] M. S. Masdar, Dedikarni, A. M. Zainoodin, M. I. Rosli, S. K. Kamarudin, and W. R. W. Daud, "Performance and stability of single and 6-cell stack passive direct methanol fuel cell (DMFC) for long-term operation," *Int. J. Hydrogen Energy*, vol. 42, no. 14, pp. 9230–9242, Apr. 2017.
[33] N. Kimiaie et al., "Results of a 20 000 h lifetime test of a 7 kW direct methanol fuel cell (DMFC) hybrid system—degradation of the DMFC stack and the energy storage," *Energy Environ. Sci.*, vol. 7, no. 9, pp. 3013–3025, Aug. 2014.
[34] A. M. I. Hassan, H. M. Saleh, and A. I. Hassan, "Synthesis and characterization of nanomaterials for application in cost-effective electrochemical devices," *Sustainability*, vol. 15, no. 14, p. 10891, Jul. 2023.
[35] Y. Wang, D. F. Ruiz Diaz, K. S. Chen, Z. Wang, and X. C. Adroher, "Materials, technological status, and fundamentals of PEM fuel cells—A review," *Mater. Today*, vol. 32, pp. 178–203, Jan. 2020.

[36] J. Ma et al ., "Killing two birds with one stone: Upgrading organic compounds via electrooxidation in electricity-input mode and electricity-output mode," *Materials*, vol. 16, no. 6, p. 2500, Mar. 2023.
[37] Z. Gao, Q. Meng, Y. Zhang, F. Zhang, Y. Luo, and H. W. Li, "An alternative for the anode materials of nickel metal hydride batteries: An AB3-type La0.6Gd0.2Mg0.2Ni2.6Co0.3Al0.1 hydrogen storage alloy," *Dalt. Trans.*, vol. 49, no. 19, pp. 6312–6320, May 2020.
[38] W. Chen, Y. Jin, J. Zhao, N. Liu, and Y. Cui, "Nickel-hydrogen batteries for large-scale energy storage," *Proc. Natl. Acad. Sci. U. S. A.*, vol. 115, no. 46, pp. 11694–11699, Nov. 2018.
[39] J. I. Martins, M. C. Nunes, R. Koch, L. Martins, and M. Bazzaoui, "Electrochemical oxidation of borohydride on platinum electrodes: The influence of thiourea in direct fuel cells," *Electrochim. Acta*, vol. 52, no. 23, pp. 6443–6449, Jul. 2007.
[40] C. Celik, F. G. Boyaci San, and H. I. Sarac, "Improving the direct borohydride fuel cell performance with thiourea as the additive in the sodium borohydride solution," *Int. J. Hydrogen Energy*, vol. 35, no. 16, pp. 8678–8682, Aug. 2010.
[41] S. Wang, A. Lu, and C. J. Zhong, "Hydrogen production from water electrolysis: Role of catalysts," *Nano Converg.*, vol. 8, no. 1, pp. 1–23, Dec. 2021.
[42] C. J. Mueller, W. J. Cannella, and G. T. Kalghatgi, "Fuels for engines and the impact of fuel composition on engine performance," *Encycl. Automot. Eng.*, pp. 1–27, Apr. 2014.
[43] G. Centi and S. Perathoner, "Catalytic technologies for the conversion and reuse of CO2," *Handb. Clim. Chang. Mitig. Adapt.*, pp. 1–50, 2021.
[44] B. C. Ong, S. K. Kamarudin, and S. Basri, "Direct liquid fuel cells: A review," *Int. J. Hydrogen Energy*, vol. 42, no. 15, pp. 10142–10157, Apr. 2017.
[45] N. S. and T. Harris, "Direct methanol fuel cells," *Encycl. Sustain. Technol.*, vol. 3, p. 15, 2017.
[46] W. C. Chiang, Y. Li, J. Shang, and T. L. Urban, "Impact of drone delivery on sustainability and cost: Realizing the UAV potential through vehicle routing optimization," *Appl. Energy*, vol. 242, pp. 1164–1175, May 2019.
[47] M. H. de Sá, C. S. Moreira, A. M. F. R. Pinto, and V. B. Oliveira, "Recent advances in the development of nanocatalysts for direct methanol fuel cells," *Energies*, vol. 15, no. 17, p. 6335, Aug. 2022.
[48] A. A. Ahmed, M. Al Labadidi, A. T. Hamada, and M. F. Orhan, "Design and utilization of a direct methanol fuel cell," *Membranes (Basel)*, vol. 12, no. 12, p. 1266, Dec. 2022.

Index

P

R

S

T